Springer Texts in Electrical Engineering

Frank M. Callier
Charles A. Desoer

Multivariable Feedback Systems

Consulting Editor: John B. Thomas

With 20 Illustrations

Springer-Verlag
New York Heidelberg Berlin

A Dowden &
Culver Book

Frank M. Callier
Department of Mathematics
Facultés Universitaires
Notre-Dame de la Paix
Namur,
Belgium

Charles A. Desoer
Department of Electrical Engineering
and Computer Sciences
University of California
Berkeley, California 94720
U.S.A.

Library of Congress Cataloging in Publication Data
Callier, F. M.
Multivariable feedback systems.
(Springer texts in electrical engineering)
Includes index.
1. Feedback control systems. I. Desoer,
Charles A. II. Title. III. Series.
TJ216.C34 1982 629.8′3 82-10407

9 8 7 6 5 4 3 2 1

ISBN-13: 978-0-387-90759-8 e-ISBN-13: 978-1-4612-5762-2
DOI: 10.1007/978-1-4612-5762-2

Preface

This volume is the result of our teaching in the last few years of a first-year graduate course on multivariable feedback systems addressed to control engineers. The prerequisites are modest: an undergraduate course in control (for acquaintance with concepts, terms, and design goals) and a senior-graduate course in linear systems. This volume covers lumped linear time-invariant multi-input multi-output systems with strong emphasis on control problems. The purpose is to provide a rapid introduction to some of the main and simpler results of control theory and to provide access to the current literature. Note that our exposition pays particular attention to the time-domain behavior of the systems under study. Note also that we cover neither optimization nor stochastic systems since these topics are treated in separate courses.

As is obvious from its abundant literature, multivariable control is a very rapidly developing field. Consequently, we have no expectation that our exposition will become definitive; however, we hope that our efforts will be found useful.

To get an idea of the contents, we suggest reading carefully the table of contents and the introduction of the chapters. Roughly, Chapter 1 is an introduction to feedback issues in a multivariable context (desensitization, large gain, singular values, etc.). Chapters 2 and 3 cover the mathematical tools for handling transfer functions as polynomial-matrix fractions and for studying systems described by polynomial matrices. Chapter 4 uses these tools to cover the general theory of interconnected systems. The last four chapters deal with feedback control theory: compensator design for controlling the dynamics, for tracking, and special design techniques for stable plants. Chapter 5 covers some of these topics in the simple single-input single-output context. The first two appendices briefly cover useful algebraic topics. The last appendix covers the division of a polynomial vector by a polynomial matrix.

In the course of developing this material we benefited from many discussions from coworkers and colleagues, and good questions from sharp-eyed students: we are deeply grateful to all of them.

We gratefully acknowledge the support over the years of the Belgian National Fund of Scientific Research, the National Science Foundation, the Joint Services Electronics Program, the Facultés Universitaires de Namur, and the University of California at Berkeley.

Our special thanks go to our wives Nicole and Jackie and our children Ann and Mike for their support during the long hours of writing.

We are very grateful to Doris Simpson for her expert typing.

Berkeley, August 1, 1981.

Contents

Note to the Reader

A course that would cover the essentials would consist of the following: Chapter 1; Sec. 2.3 and Secs. 2.4.1 to 2.4.4; Secs. 3.2.1, 3.2.2, 3.2.3 (results only), 3.3.1 (results), 3.3.2, and 3.4 (results); Chapter 4; Chapter 5; for the last three chapters the introductions provide guidance to the topics of interest.

We have not hesitated to lighten our style by using a number of mathematical symbols, e.g., $\mathbb{C}_+$ for the closed right-half of the complex plane; these symbols are almost standard in the literature. For convenience a table of symbols is placed ahead of the subject index (p. 264).

We also use a number of abbreviations, e.g., PMD for polynomial matrix system description; they are included at their lexicographic locations in the subject index.

Chapter 1. On the Advantages of Feedback

1.1. Introduction

Feedback is a major engineering invention. The main reasons for using feedback are:

1. Desensitization of the closed-loop system performance due to plant variations (see Sec. 1.3).
2. Reduction of the closed-loop system response due to external disturbances or noise (see Sec. 1.3).
3. Robust asymptotic tracking and robust disturbance rejection for certain classes of inputs (see Sec. 5.2 and 7.2).
4. Improvement of dynamic response: e.g., stabilization of unstable or insufficiently stable plants (see Secs. 5.2 and 6.2 and Chapter 8).
5. Achieving a closed-loop system that is more linear than the plant (e.g., hi-fi amplifiers, measuring equipment, op-amps).
6. Positive feedback is used to create instability (e.g., oscillators, flip-flops, Q-multipliers).

This chapter investigates the conditions under which large loop gain achieves feedback objectives 1 and 2.

For multiple-feedback-loop systems, the notion of large loop gain needs a clear definition. It turns out that the singular value decomposition of a matrix is an appropriate tool for describing this notion.

Thus Sec. 1.2 describes the singular value decomposition of a complex matrix (Theorem 1.2.21), and develops the idea of directional gain (Theorem 1.2.57).

Section 1.3 studies closed-loop exponentially stable feedback systems in the frequency domain under large loop gain. The notion of large loop gain is defined and its effects are investigated (a) for response shaping and desensitization of the closed-loop output under external disturbances (Theorem 1.3.24), and (b) for desensitization under plant variations (Theorems 1.3.35 and 1.3.43).

Although the analysis in Sec. 1.3 is carried out algebraically in the frequency domain, there is everywhere an implied time-domain interpretation. In particular, with $\mathcal{R}(0)$, viz., the ring of exponentially stable transfer functions, given by

$$R(0) := \{f \in \mathbb{R}_p(s) : f \text{ is analytic in } \mathbb{C}_+\},$$

closed-loop exponential stability will mean here that all closed-loop transfer functions have entries in $R(0)$, or equivalently are exponentially stable. We shall then assume that all open-loop transfer functions are generated by an underlying time-domain model, called PMD (see Sec. 3.2), which is well-formed and has no unstable hidden modes (see Sec. 3.3). Consequently, the notion of closed-loop exponential stability will coincide with the notion of exponential stability of the underlying closed-loop PMD (see Secs. 3.3 and 4.2): the latter notion has time-domain implications given by Theorem 3.3.2.19.

1.2. Singular Value Decomposition of a Matrix

Singular value decomposition (s.v.d.) is an algorithm which determines the singular values of a matrix $A \in \mathbb{C}^{m\times n}$. The singular values allow one to determine the size of the action of the underlying linear map in all directions. As a consequence it will be possible to express mathematically that the action is small or large in all directions ("small" or "large" gain).

The exercises that follow delineate useful properties of any matrix $A \in \mathbb{C}^{m\times n}$.

1 Exercise [Orthogonal decomposition of domain and codomain]

Show that if $A \in \mathbb{C}^{m\times n}$ with rank r,

2 $\quad \mathbb{C}^n = \text{dom}(A) = R(A^*) \overset{\perp}{\oplus} N(A),$

3 $\quad \mathbb{C}^m = \text{codom}(A) = R(A) \overset{\perp}{\oplus} N(A^*),$

4 $\quad \text{rk } A = \text{rk } A^* = r.$

5 Exercise [Bijections induced by orthogonal decomposition]

Show that if $A \in \mathbb{C}^{m\times n}$ with rank r, then with $A|_{R(A^*)}$ and $A^*|_{R(A)}$ denoting the restrictions of A on $R(A^*)$ and A^* on $R(A)$, resp.:

6 $\quad A|_{R(A^*)} : R(A^*) \to R(A)$ is a bijection,

7 $\quad A^*|_{R(A)} : R(A) \to R(A^*)$ is a bijection,

8 $\quad R(A^*) = R(A^*A)$ and $N(A) = N(A^*A)$,

9 $\quad R(A) = R(AA^*)$ and $N(A^*) = N(AA^*)$,

10 $\quad \mathrm{rk}(AA^*) = \mathrm{rk}(A) = \mathrm{rk}(A^*) = \mathrm{rk}(A^*A) = r.$

11 Comment. Exercises 1 and 5 show that any linear map represented by a matrix $A \in \mathbb{C}^{m\times n}$ can be made invertible by restricting its domain and codomain. The same holds for $A^* \in \mathbb{C}^{n\times m}$. The situation is displayed by Fig. 1.

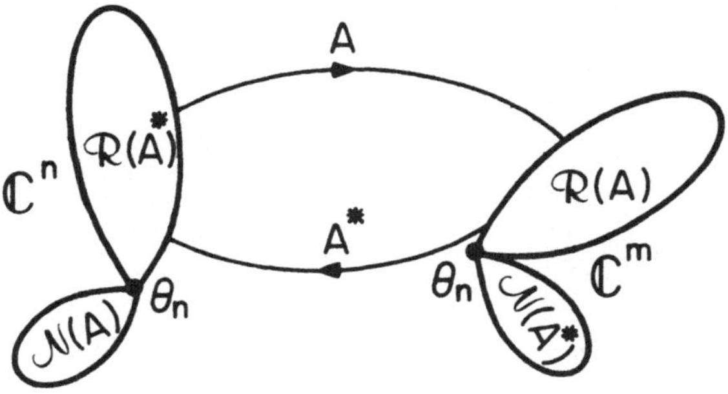

Fig. 1. Bijections associated with $A \in \mathbb{C}^{m\times n}$ through orthogonal decomposition of domain and codomain.

S.v.d. will display bijections (6) and (7) in a simple diagonal form by the choice of orthonormal bases for $\mathcal{R}(A^*)$ and $\mathcal{R}(A)$. We note further that (a) maps (6) and (7) are adjoints of each other (indeed, $\forall x \in \mathcal{R}(A^*)$ and $\forall y \in \mathcal{R}(A)$ $\langle Ax,y\rangle = \langle x,A^*y\rangle$); and (b) the composition of maps (6) and (7) delivers two bijections $A^*A\big|_{\mathcal{R}(A^*)} : \mathcal{R}(A^*) \to \mathcal{R}(A^*)$ and $AA^*\big|_{\mathcal{R}(A)} : \mathcal{R}(A) \to \mathcal{R}(A)$ which are Hermitian positive definite, having as eigenvalues the nonzero eigenvalues of resp. A^*A and AA^*. We have also

15 Exercise [Common nonzero eigenvalues of A^*A and AA^*]. Let $A \in \mathbb{C}^{m\times n}$. Show that for the Hermitian positive semidefinite matrices $A^*A \in \mathbb{C}^{n\times n}$ and $AA^* \in \mathbb{C}^{m\times m}$, $\forall$ nonzero $\sigma \in \mathbb{R}$,

$$16 \qquad \det\begin{array}{c} \begin{matrix} m & \; n \end{matrix} \\ \left[\begin{array}{c|c} \sigma I_m & A \\ \hline A^* & \sigma I_n \end{array}\right]\begin{matrix} m \\ n \end{matrix} \end{array} = \sigma^{m-n}\det[\sigma^2 I_n - A^*A] = \sigma^{n-m}\det[\sigma^2 I_m - AA^*],$$

17 A^*A and AA^* have identical nonzero eigenvalues $\sigma^2 > 0$.

18 Comment. The s.v.d. of $A \in \mathbb{C}^{m\times n}$ will display the square roots of these eigenvalues: these are called positive singular values of A.

21 Theorem [Singular value decomposition of a complex matrix]. Let $A \in \mathbb{C}^{m\times n}$ be of rank r. Then there exist matrices $U \in \mathbb{C}^{m\times m}$, $V \in \mathbb{C}^{n\times n}$, $\Sigma_1 \in \mathbb{R}^{r\times r}$ s.t.

(a)

22 $$V = \underset{}{\left[\, \underset{r}{V_1} \;\vdots\; \underset{n-r}{V_2} \,\right]_n} \in \mathbb{C}^{n\times n}$$

satisfies

V is unitary, i.e., $V^*V = I_n$,

$\mathcal{R}(V_1) = \mathcal{R}(A^*)$,

The columns of V_1 form an orthonormal basis for $\mathcal{R}(A^*)$,

$\mathcal{R}(V_2) = \mathcal{N}(A)$,

The columns of V_2 form an orthonormal basis for $\mathcal{N}(A)$,

The columns of V form a complete orthonormal basis of eigenvectors of A^*A.

(b)

23 $$U = \left[\, \underset{r}{U_1} \;\vdots\; \underset{m-r}{U_2} \,\right]_m \in \mathbb{C}^{m\times m}$$

satisfies

U is unitary, i.e., $U^*U = I_m$,

$\mathcal{R}(U_1) = \mathcal{R}(A)$,

The columns of U_1 form an orthonormal basis for $\mathcal{R}(A)$,

$\mathcal{R}(U_2) = \mathcal{N}(A^*)$,

The columns of U_2 form an orthonormal basis for $\mathcal{N}(A^*)$,

The columns of U form a complete orthonormal basis of eigenvectors of AA^*.

(c) Under the vector representations for $\mathcal{R}(A^*)$ and $\mathcal{R}(A)$ given by

24 $$\mathbb{C}^r \to \mathcal{R}(A^*) : \xi^1 \mapsto x^1 = V_1\xi^1, \quad \mathbb{C}^r \to \mathcal{R}(A) : \eta^1 \mapsto y^1 = U_1\eta^1,$$

the bijections induced by orthogonal decomposition, viz.,

25 $$A\big|_{R(A^*)} : R(A^*) \to R(A), \quad A^*\big|_{R(A)} : R(A) \to R(A^*)$$

have representations

26 $$\Sigma_1 : \xi^1 \mapsto \eta^1 = \Sigma_1 \xi^1, \quad \Sigma_1 : \eta^1 \mapsto \xi^1 = \Sigma_1 \eta^1,$$

with

27 $$\Sigma_1 := \operatorname{diag}[\sigma_1, \sigma_2, \cdots, \sigma_r] \in \mathbb{R}^{r\times r},$$

s.t. $$\sigma_1 \geq \sigma_2 \geq \cdots \geq \sigma_r > 0,$$

where the σ_i, $i \in \underline{r}$, are the square roots of the common nonzero eigenvalues of A^*A and AA^*, called <u>positive singular values of A</u>.

(d) $A \in \mathbb{C}^{m\times n}$ has a <u>dyadic expansion</u>

30 $$A = U_1\Sigma_1 V_1^* \text{ or equiv. } A = \sum_{i=1}^{r} \sigma_i u_i v_i^*$$

where the u_i, v_i for $i \in \underline{r}$ are columns of U_1, resp. V_1.

(e) $A \in \mathbb{C}^{m\times n}$ has a <u>singular value decomposition</u> (s.v.d.)

31 $$A = U\,\Sigma\, V^*,$$

where

32 $$\Sigma := \begin{bmatrix} \Sigma_1 & 0 \\ 0 & 0 \end{bmatrix} \begin{matrix} r \\ m-r \end{matrix}, \quad \text{columns: } r,\ n-r$$ ∎

37 <u>Comments</u>. (a) If $A \in \mathbb{R}^{m\times n}$, then $U \in \mathbb{R}^{m\times m}$ and $V \in \mathbb{R}^{n\times n}$ are orthogonal ($U^TU = I_m$, etc.) and define "rotations" on codomain and domain.

(b) Theorem 21 teaches us that, modulo unitary transformations in the domain and codomain, matrix $A \in \mathbb{C}^{m\times n}$ can be diagonalized displaying the bijections induced by orthogonal decomposition of Exercise 5 in their simplest diagonal form: the positive singular values so displayed are the square roots of the nonzero eigenvalues of both AA^* and A^*A (see Comments 11 and 18).

(c) Appropriate references for the theory are [Ste.1]; for programs and subroutines, see [Gar.1].

(d) S.v.d. has the following geometric content: from (22), (23), and (31), $A[V_1 \vdots V_2] = [U_1\Sigma_1; 0]$, whence for the "active part" $AV_1 = U_1\Sigma_1$, i.e., $Av_i = u_i\sigma_i$ for $i \in \underline{r}$: <u>any linear map "rotates"</u> ($v_i \mapsto u_i$, $i \in \underline{r}$) <u>and "scales"</u> ($u_i \mapsto u_i\sigma_i$, $i \in \underline{r}$) (the v_i's and u_i's are columns of V_1 and U_1).

40 <u>Proof of Theorem 21</u>. By construction.

(a) $A \in \mathbb{C}^{m\times n}$ has rank r, whence by (10) the Hermitian p.s.d. matrix A^*A has rank r with n nonnegative eigenvalues σ_i^2, $i \in \underline{n}$, ordered as

41 $$\sigma_1^2 \geq \sigma_2^2 \geq \cdots \geq \sigma_r^2 > 0 = \sigma_{r+1}^2 = \cdots = \sigma_n^2,$$

to which correspond a complete orthonormal eigenvector basis $(v_i)_1^n$ of A^*A. This ordered set of $\mathbb{C}^n$ vectors defines a unitary $n \times n$ matrix V having a partition and properties (22) using (8).

(b) The $m \times m$ matrix U is now constructed as follows. Define

42 $$\Sigma_1 := \text{diag}[\sigma_1, \sigma_2, \cdots, \sigma_r],$$

where the σ_i, $i \in \underline{r}$, are those given in (41). From properties (22), especially V_1 is a complete set of orthonormal eigenvector for the nonzero eigenvalues of A^*A, $A^*AV_1 = V_1\Sigma_1^2$, whence $(AV_1\Sigma_1^{-1})^*(AV_1\Sigma_1^{-1}) = I_r$. This defines a $m \times r$ matrix

43 $$U_1 := AV_1\Sigma_1^{-1}$$

and from the above we obtain by calculation

$$U_1^*U_1 = I_r$$

$$AA^*U_1 = U_1\Sigma_1^2.$$

Since A^*A and AA^* have identical nonzero eigenvalues (hence the diagonal elements of Σ_1^2 include <u>all</u> the positive eigenvalues of AA^*), it follows that the columns of U_1 form an orthonormal basis for $R(AA^*) = R(A)$: properties (23^2) and (23^3) hold.

Define now any $m \times (m-r)$ matrix U_2 with <u>orthonormal</u> columns s.t. $U_2^*U_1 = 0$. Note that

$$U = \begin{bmatrix} \underset{r}{U_1} & \vdots & \underset{m-r}{U_2} \end{bmatrix} m \in \mathbb{C}^{m\times m}$$

is a unitary matrix confirming (23^1); moreover, $\mathbb{C}^m = R(U_1) \overset{\perp}{\oplus} R(U_2)$. Now by (3), $\mathbb{C}^m = R(A) \overset{\perp}{\oplus} N(A^*)$, where we know that $R(U_1) = R(A)$: hence $R(U_2) = N(A^*)$ and (23^4) and (23^5) follow. Notice also that by (9) $N(A^*) = N(AA^*)$. Therefore, property (23^6) follows since U is unitary, $AA^*U_1 = U_1\Sigma_1^2$ and $R(U_2) = N(AA^*)$.

(c) The proof of assertion (c) uses the relations

44 $$AV_1 = U_1\Sigma_1 \qquad A^*U_1 = V_1\Sigma_1$$

which follow from (43); it also uses the fact that the nonzero eigenvalues of A*A in (41)-(42) are also those of AA*.

(d) The dyadic expansion (30) follows from (44).

(e) The singular value decomposition (31)-(32) follows because

$$A[V_1 \,\vdots\, V_2] = [U_1\Sigma_1 \,\vdots\, 0] = [U_1 \,\vdots\, U_2]\Sigma.$$ ∎

For motivating the definition and results below, we have

48 Exercise. Show that a unitary transformation $x = V\xi$ (with $V^*V = I_n$) leaves the Euclidean norm unchanged, or equiv. $\|x\| = \|\xi\|$.

49 Exercise. Let $A \in \mathbb{C}^{m\times n}$ have a s.v.d. (31)-(32) and apply the unitary transformations $x = V\xi$, $y = U\eta$, where V and U are given by (22) and (23). Then

50 $$\|Ax\| = \|\Sigma\xi\|.$$

Hence for $x = v_i$ (hence $\xi = (0, \cdots, 0, 1, 0, \cdots, 0)^T$) with $i \in \underline{n}$

51 $$\|Av_i\| = \sigma_i \qquad \forall i \in \underline{n},$$

where the σ_i are the square roots of the n eigenvalues of A*A.

52 Comment. Note that the n square roots of the eigenvalues σ_i^2 of A*A measure the size of the actions of A on a complete set of orthonormal directions of the domain of A. For this reason we have

53 Definition. Let $A \in \mathbb{C}^{m\times n}$ have rank r. Then the n nonnegative square roots σ_i of the eigenvalues of A*A are called singular values of A. When ordered according to

$$\sigma_1 \geq \sigma_2 \geq \cdots \geq \sigma_2 > 0 = \sigma_{r+1} = \cdots = \sigma_n$$

the first r singular values are called <u>positive</u> and are given by the s.v.d. (31)-(32).

A complete picture of the action of a matrix $A \in \mathbb{C}^{m\times n}$ is the image under A of the unit sphere of $\mathbb{C}^n$. In fact, this combines display of <u>directional influence</u> and <u>size of action</u>.

57 <u>Theorem</u> [Indicator ellipsoid of a matrix $A \in \mathbb{C}^{m\times n}$]. Let $A \in \mathbb{C}^{m\times n}$ be of rank r and have a s.v.d. (31)-(32). Consider the domain and codomain coordinate transformations defined by $x = V\xi$ and $y = U\eta$ with U and V defined by (22) and (23).

Let S_n denote the unit sphere of $\mathbb{C}^n$ centered at θ_n, equiv.

58 $$S_n := \{x \in \mathbb{C}^n : \|x\| = 1\}.$$

Let $A[S_n]$ denote the image under A of S_n. U.t.c. with $\eta = (\eta_i)_1^m \in \mathbb{C}^m$,

59 $$r = m = n \Rightarrow A[S_n] = \{y \in \mathbb{C}^m : y = U\eta, \sum_{i=1}^{r} (\eta_i/\sigma_i)^2 = 1\},$$

60 $$r = m < n \Rightarrow A[S_n] = \{y \in \mathbb{C}^m : y = U\eta, \sum_{i=1}^{r} (\eta_i/\sigma_i)^2 \leq 1\},$$

61 $$r = n < m \Rightarrow A[S_n] = \{y \in \mathbb{C}^m : y = U\eta, \sum_{i=1}^{r} (\eta_i/\sigma_i)^2 = 1, \eta_{r+1} = \cdots = \eta_m = 0\},$$

62 $$r < \min(m,n) \Rightarrow A[S_n] = \{y \in \mathbb{C}^m : y = U\eta, \sum_{i=1}^{r} (\eta_i/\sigma_i)^2 \leq 1, \eta_{r+1} = \cdots = \eta_m = 0\}.$$

63 <u>Comment</u>. A maps the <u>unit sphere of $\mathbb{C}^n$ onto an r-dimensional ellipsoid in $R(A)$</u> with as principal axes the columns u_i of U of length $\sigma_i > 0$ for $i \in \underline{r}$: the points in $R(A)$ interior to the ellipsoid have to be included iff A has not full column rank: the ellipsoid has m principal axes with positive length iff A has full row rank. Note that the four cases in (59)-(62) are <u>mutually exclusive</u>. Notice also that <u>lengths of the principal axes</u> are the <u>positive singular values</u> and that the relationship between the principal axes for S_n and the ellipsoid is

$$Av_i = \sigma_i u_i \qquad i \in \underline{r},$$

where the u_i and v_i are columns of U_1 and V_1: see the s.v.d. of A.

66 Proof of Theorem 57. We prove only the completely degenerate case $r < \min(m,n)$. Since $\mathbb{C}^n = R(A^*) \overset{\perp}{\oplus} N(A)$, it follows that $A[S_n]$ is the image under A of the orthogonal projection of S_n onto $R(A^*)$ which is B_r, viz., the unit ball of $R(A^*)$ centered at θ_n. Under the coordinate transformation $x = V\xi$, where V is given by (22), we have, with $\xi = (\xi_i)_{i=1}^n$,

$$B_r = \{x \in \mathbb{C}^n : x = V\xi, \sum_{i=1}^{r} \xi_i^2 \leq 1, \xi_{r+1} = \cdots = \xi_n = 0\}$$

and

$$A[S_n] = A[B_r] \subset R(A).$$

Now using the s.v.d. $AV = U\Sigma$ and $y = U\eta$ with U given by (23), it follows that $\eta = \Sigma\xi$ with $\xi_i = \eta_i/\sigma_i$ for $i \in \underline{r}$ and $\eta_{r+1} = \cdots = \eta_m = 0$: (62) holds. ∎

67 Exercise. Prove assertions (59)-(61) of Theorem 57.

We are now in a position to discuss norm implications. Note that for any matrix $A \in \mathbb{C}^{m\times n}$ the ℓ_2-induced norm of A is given by

68 $$\|A\| = \max_{x\neq\theta} \{\|Ax\|/\|x\|\} = \max\{\|Ax\| : \|x\| = 1\}.$$

Hence

$$\|A\| = \max\{\|y\| : y \in A[S_n]\}:$$

the norm is the size of longest vector of the image under A of the unit sphere of $\mathbb{C}^n$.

71 Theorem [Norm of a matrix and of its inverse]. Let $A \in \mathbb{C}^{m\times n}$ be of rank r and have a s.v.d. (31)-(32). In particular, A has n singular values

72 $$\sigma_1 \geq \sigma_2 \geq \cdots \geq \sigma_r > 0 = \sigma_{r+1} = \cdots = \sigma_n.$$

U.t.c.

(a)

73 $$\|A\| = \max\{\|Ax\| : \|x\| = 1\} = \sigma_1$$

(b)

74 $$\min\{\|Ax\| : \|x\| = 1\} = \sigma_n,$$

where

75 $$\sigma_n > 0 \Leftrightarrow A \text{ has full column rank.}$$

(c) If A is square and nonsingular ($r = m = n$), then

76 $$\|A^{-1}\| = 1/\sigma_n.$$

77 Proof. Exercise: use Theorem 57 and the norm invariance of unitary transformations.

78 Comments. (a) The largest singular value is the norm of A and the smallest singular value is the inverse of the norm of A^{-1}.
(b) In the square nonsingular case σ_n is a worst-case sensitivity parameter for the equation $Ax = b$, $b \in \mathbb{C}^n$. Perturb b by δb and calculate δx: $\delta x = A^{-1}\delta b$. If $\delta b = \delta\alpha\, u_n$, where $\delta\alpha$ is scalar and u_n is the nth column of U of (23), then $\|\delta x\| = |\delta\alpha|\, 1/\sigma_n$ (thus when $\sigma_n \ll 1$, a small δb can cause a very large δx). Using the indicator ellipsoid of A^{-1}, it is seen that b-perturbations in any other direction cause smaller x-perturbations.
(c) If we want the size of the action of A to be large (small) in all directions, we must require σ_n to be large (σ_1 to be small resp.).

1.3. Large Loop Gain

In this section we consider a multivariable feedback system Σ given by Fig. 1 under the following assumptions.

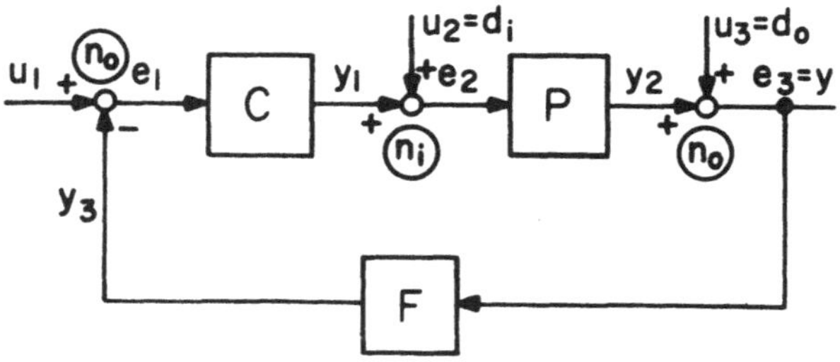

Fig. 1. Feedback system Σ.

1 Assumption 1. P is the plant and C and F are resp. the precompensator and feedback compensators described by their transfer functions

$$P \in \mathbb{R}_{p,o}(s)^{n_o \times n_i}, \quad C \in \mathbb{R}_p(s)^{n_i \times n_o}, \quad F \in \mathbb{R}_p(s)^{n_o \times n_o},$$

where it is assumed that P, C, and F have underlying well-formed PMDs with no unstable hidden modes (see Chapter 3). ■

Note that each subsystem is preceded by a summing node with exogenous inputs u_1, $u_2 = d_i$, $u_3 = d_o$, which are the closed-loop system input, and additive disturbances at the input and output of the plant; e_i and y_i for $i = 1 \sim 3$ are the subsystem input and outputs; in particular $e_3 = y$ is the closed-loop system output.

2 Assumption 2. Feedback system Σ is exp. stable, or equiv. the closed-loop transfer functions $H_{y_j u_i} : u_i \to y_j$, for i and $j = 1 \sim 3$, are exp. stable (equiv. $\in E(R(0))$; see (2.2.6)). ■

3 Comment. By the methods of Chapter 4 it can be shown that under Assumptions 1 and 2 all closed-loop transfer functions are exp. stable.

We propose now to study the frequency responses $H_{yu_1}(j\omega)$ and $H_{yd_0}(j\omega)$ (i.e., of the closed-loop output due to input and disturbance at the output of the plant) under "large loop gain." From Fig. 1 one has

7 $$H_{yu_1} = PC(I + FPC)^{-1} = (I + PCF)^{-1}PC$$

8 $$H_{yd_0} = (I + PCF)^{-1},$$

9 $$H_{yu_1} - F^{-1} = -(I + PCF)^{-1}F^{-1},$$

where in (9) we have assumed that F is invertible. We shall denote by $\|P\|$, $\sigma_{max}[P]$, $\sigma_{min}[P]$ the ℓ_2-induced norm and the maximum and minimum singular value of the complex matrix transfer function P at $j\omega$, $\omega \in \mathbb{R}$. For feedback system Σ of Fig. 1, PCF, (I + PCF), $(I + PCF)^{-1}$ are usually called the loop-, return difference- and sensitivity transfer functions.

Recall Theorem 1.2.57. Taking norms on both sides of (8) and (9), we note that if the sensitivity is small over Ω, equiv.

10 $$\|(I + PCF)^{-1}\| = \sigma_{max}[(I + PCF)^{-1}] << 1 \quad \text{over } \Omega,$$

where Ω is a bounded frequency interval (as $|\omega| \to \infty, (I + PCF)^{-1} \to I$), then

11 $$\|H_{yd_0}\| << 1 \quad \text{over } \Omega,$$

12 $$\|H_{yu_1} - F^{-1}\| << \|F^{-1}\| \quad \text{over } \Omega,$$

i.e., over Ω the closed-loop output y can be made insensitive to disturbances at the output of the plant and the I/O map: $u_1 \mapsto y$ is approximately equal to F^{-1}; i.e., we get response shaping over Ω.

Let us tie up (10) with "large loop gain."

16 Definition. We say that the m.i.m.o. feedback system Σ of Fig. 1 has a large loop gain over Ω iff

17 $$\sigma_{min}[PCF] >> 1 \quad \text{over } \Omega.$$ ∎

18 Comment. In view of the s.v.d. of PCF at $j\omega$, condition (17) requires the size of the action of $PCF(j\omega)$ to be large in all directions $\forall \omega \in \Omega$.

The connection with (10) is

19 Lemma. Given the feedback system Σ of Fig. 1. U.t.c.

20 $$\|(I + PCF)^{-1}\| = \sigma_{max}[(I + PCF)^{-1}] << 1 \quad \text{over } \Omega$$

$\Leftrightarrow$

21 $$\sigma_{min}[I + PCF] >> 1 \quad \text{over } \Omega$$

$\Leftrightarrow$

17 $$\sigma_{min}[PCF] >> 1 \quad \text{over } \Omega.$$

22 Comment. Small sensitivity over Ω means a large return difference and large loop gain over Ω.

23 Proof of Lemma 19. Use ℓ_2-norms and Theorem 1.2.57.

(20) $\Leftrightarrow$ (21): $\sigma_{max}[(I + PCF)^{-1}] = \|(I + PCF)^{-1}\| = (\sigma_{min}[I + PCF])^{-1}$.

(21) $\Leftrightarrow$ (17): at $\omega \in \Omega$, $\forall x \in \mathbb{C}^{n_o}$ s.t. $\|x\| = 1$

$$\|PCFx\| - \|x\| \le \|(I + PCF)x\| \le \|PCFx\| + \|x\|.$$

Taking successive minima over $x \in \mathbb{C}^{n_o}$ with $\|x\| = 1$,

$$\sigma_{min}[PCF] - 1 \le \sigma_{min}[I + PCF] \le \sigma_{min}[PCF] + 1.$$

Hence

$$\sigma_{min}[PCF] >> 1 \;\Rightarrow\; \sigma_{min}[I + PCF] >> 1 \;\Rightarrow\; \sigma_{min}[PCF] >> 1.$$ ∎

Hence by (10)-(12) and Lemma 19 one has

24 Theorem. Given feedback system Σ of Fig. 1. U.t.c.

17 $\sigma_{min}[PCF] >> 1$ over Ω

$\Rightarrow$

11 $\|H_{yd_0}\| << 1$ over Ω

12 $\|H_{yu_1} - F^{-1}\| << \|F^{-1}\|$ over Ω.

25 Comment. If the loop gain is large over Ω, the closed-loop output y can be made insensitive to disturbances applied at the plant output and the I/O-map is approximately equal to F^{-1}: for F = I we get approximate decoupling over frequency band Ω.

We study next the sensitivity of the closed-loop response w.r.t. plant variations.

28 Assumption 3. Considering Fig. 1, the nominal plant $P \in \mathbb{R}_{p,o}(s)^{n_o \times n_i}$ is replaced by a perturbed plant $\tilde{P} \in \mathbb{R}_{p,o}(s)^{n_o \times n_i}$ such that (a) the nominal feedback system Σ is replaced by a perturbed feedback system $\tilde{\Sigma}$ and (b) Assumptions 1 and 2 are still valid. ∎

For the feedback configuration of Fig. 1 a plant variation

29 $$\Delta P = \tilde{P} - P$$

generates a variation of the closed-loop I/O map given by

30 $$\Delta H_{yu_1} = \tilde{H}_{yu_1} - H_{yu_1}.$$

By equation (7), the I/O map H_{yu_1} can also be obtained as the I/O map of the <u>nominally equivalent open-loop system</u> [Cru.1] shown in Fig. 2. For this open-loop configuration a plant variation (29) generates

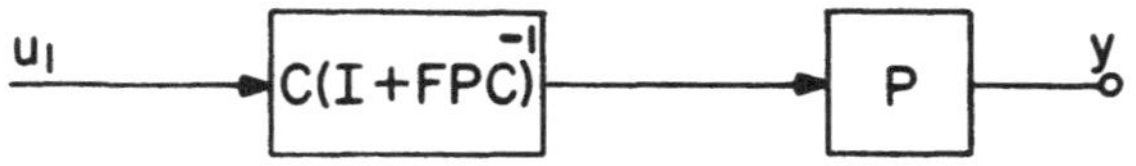

Fig. 2. The nominally equivalent open-loop system.

an I/O map variation

31 $$\Delta H^0_{yu_1} = \Delta P \cdot C(I + FPC)^{-1}.$$

Using (7) and (29)-(31) ΔH_{yu_1}, the variation of the closed-loop I/O map, is given in terms of $\Delta H^0_{yu_1}$, the variation in terms of the nominally equivalent open-loop system, by

32 $$\begin{aligned}\Delta H_{yu_1} &= (I + \tilde{P}CF)^{-1}\tilde{P}C - PC(I + FPC)^{-1} \\ &= (I + \tilde{P}CF)^{-1}\Delta PC(I + FPC)^{-1} = (I + \tilde{P}CF)^{-1}\Delta H^0_{yu_1}.\end{aligned}$$

We now have

35 <u>Theorem</u>. Consider the feedback system of Fig. 1 and the nominally equivalent open-loop system of Fig. 2 subject to plant variations (29) s.t. Assumption 3 holds.

U.t.c.

The closed-loop and nominally equivalent open-loop variations in the input-output map (30) and (31) are related by

36 $$\Delta H_{yu_1} = (I + \tilde{P}CF)^{-1}\Delta H^0_{yu_1}.$$

Therefore, if

37 $$\sigma_{min}[\tilde{P}CF] >> 1 \qquad \text{over } \Omega,$$

then

38 $$\|\Delta H_{yu_1}\| << \|\Delta H^0_{yu_1}\| \qquad \text{over } \Omega.$$ ⌘

39 Comment. <u>If the loop gain of the perturbed system is large over Ω, then the closed-loop I/O map variation is much smaller than that of the nominally equivalent open-loop system</u>.

42 Proof of Theorem 35. (36) has been obtained in (32). Taking norms at $\omega \in \Omega$, (36) gives

$$\|\Delta H_{yu_1}\| \leq \|(I + \tilde{P}CF)^{-1}\| \|\Delta H^0_{yu_1}\|.$$

Therefore, (37) implies (38) by Lemma 19. ⌘

The following Theorem compares "percentage changes" for the case of a <u>square</u> invertible plant [Saf.1].

43 Theorem. Consider the feedback system of Fig. 1 with $n_o = n_i$. Consider the variation of the closed-loop input-output map (30) generated by a plant variation (29) s.t. Assumption 3 holds.

U.t.c.

(a)

44 $$\Delta H_{yu_1} \cdot H^{-1}_{yu_1} = (I + \tilde{P}CF)^{-1}(\Delta P \cdot P^{-1}).$$

Consequently, if

45 $$\sigma_{min}[\tilde{P}CF] >> 1 \qquad \text{over } \Omega,$$

then

46 $$\|\Delta H_{yu_1} \cdot H^{-1}_{yu_1}\| << \|\Delta P \cdot P^{-1}\| \qquad \text{over } \Omega.$$

(b)

47 $$\Delta H_{yu_1} \cdot \tilde{H}^{-1}_{yu_1} = (I + PCF)^{-1}(\Delta P \cdot \tilde{P}^{-1}).$$

Consequently, if

17 $$\sigma_{min}[PCF] \gg 1 \qquad \text{over } \Omega,$$

then

48 $$\|\Delta H_{yu_1} \cdot \tilde{H}_{yu_1}^{-1}\| \ll \|\Delta P \cdot \tilde{P}^{-1}\| \qquad \text{over } \Omega,$$ ∎

49 *Comment*. Formulas (44) and (47) relate *relative* changes in the closed-loop I/O map to *relative* change in the plant. These are w.r.t. nominal values in (44) and w.r.t. perturbed values in (47) while the reverse takes place for the sensitivities. The results are similar: *for a large loop gain over Ω, the I/O map is made insensitive to plant variations*.

50 *Proof of Theorem 43*. By (32) and (7) one obtains (44); (46) is obtained taking norms and Lemma 19.

51 *Exercise*. Prove assertions (47) and (48).

52 *Conclusion*. For the m.i.m.o. feedback system of Fig. 1 a large loop gain ensures

(1) that the closed-loop I/O map is approximately F^{-1} (roughly independent of C and P);

(2) desensitization of the closed-loop system output y due to disturbances applied at the output of the plant;

(3) densensitization of the closed-loop I/O map due to plant variations. ∎

53 *Comment*. Since F determines the I/O map over Ω (see (12)), it is the precompensator C that has to be made large in order to obtain a large loop gain PCF. It is well known that, in most cases, if the loop gain is made too large, the system becomes unstable. Also, if C becomes large, it will amplify any noise present at the input summing node and may cause plant saturation.

54 *Comment*. For many uses of singular values, see IEEE Trans. AC-26, No. 1, Feb. 1981. Some aspects of the *nonlinear* theory of large loop gain may be found in [Des.3], [Saf.1], and [Cru.1].

Chapter 2. Matrix Fraction Description of Transfer Functions

2.1. Introduction

We all know the state space description of linear lumped time-invariant systems:

1a $$\dot{x}(t) = Ax(t) + Bu(t)$$
$$t \geq 0$$
1b $$y(t) = Cx(t) + Du(t)$$

where A, B, C, D are constant matrices and $x(\cdot)$, $y(\cdot)$ and $u(\cdot)$ are, respectively, the state, the output, and the input of the system. With $p := d/dt$, these equations read

2a $$(pI-A)x(t) = Bu(t)$$

2b $$y(t) = Cx(t) + Du(t).$$

We may view these equations as specified by a quadruple of polynomial matrices $[pI - A, B, C, D]$. More generally we may consider a quadruple of polynomial matrices: $[D, N_\ell, N_r, K]$ and specify the system by

3a $$D(p)\xi(t) = N_\ell(p)u(t)$$
$$t \geq 0$$
3b $$y(t) = N_r(p)\xi(t) + K(p)u(t)$$

This is called a polynomial matrix system description (PMD) [Ros.1,Kai.1]. (For a precise definition, see Sec. 3.2.1.).

The PMD described by (3) has the transfer function

4 $$H = N_r D^{-1} N_\ell + K \in \mathbb{R}(s)^{n_o \times n_i}$$

In many physical problems, the linearized equations lead to expressions of the form

5 $$D_\ell(p)\xi(t) = N_\ell(p)u(t) \qquad y(t) = \xi(t),$$

with the transfer function

6 $$H = D_\ell^{-1}N_\ell \in \mathbb{R}(s)^{n_o \times n_i}.$$

In circuit theory, for example, the tableau equations lead to equations of the form

7 $$D_r(p)\xi(t) = u(t)$$
$$y(t) = N_r(p)\xi(t) + K(p)u(t)$$

with transfer function

8 $$H = N_rD_r^{-1} + K = (N_r + KD_r)D_r^{-1}$$

Equation (6) gives a left matrix fraction of H and (8) gives a right matrix fraction of H.

It is for this reason that we study in this chapter polynomial matrix fractions of rational matrices. Section 2 develops the algebraic tools for studying polynomial matrices (the Euclidean ring of polynomials, modules of polynomial vectors, ···). Section 3 studies polynomial matrices: first, algebraically, (coprimeness, elementary operations, greatest common divisors, ···), and then dynamically (the differential equation $D(p)\xi(t) = \theta$, ···). Section 4 studies polynomial matrix fractions: first algebraically, (coprime and proper fractions, division ···) and then the related system theoretic properties (poles and zeros, dynamical interpretations, ···). Section 5 sketches how a matrix fraction transfer function can be realized as a state space system.

2.2. Polynomials, Euclidean Rings, and Modules

In this section we describe some algebraic structures and facts related to polynomials and polynomial matrices. The reader is assumed familiar with the algebraic appendices A and B.

1 Definition. A Euclidean ring $(R, +, \cdot, 0, 1)$ is a commutative entire ring, (A.10), upon which there is defined a gauge [Sig. 1,p.132], i.e., a function

$$\gamma : R\backslash\{0\} \to \mathbb{N} : a \mapsto \gamma(a)$$

s.t. the following axioms hold:

ER1. $\forall a, b \in R\backslash\{0\}$ $\gamma(a) \le \gamma(ab)$

ER2. If $a \in R$, $b \in R\backslash\{0\}$, then $\exists$ elements q and r in R s.t.
$a = bq + r$, where $r = 0$ or $\gamma(r) < \gamma(b)$. ∎

2 Comment. Roughly speaking, a Euclidean ring is an entire ring in which a division operation is defined and which delivers a quotient q and a remainder r: ER2 is called the Euclidean division property.

3 Fact. $(\mathbb{R}[s], +, \cdot, 0, 1)$ is a Euclidean ring under pointwise addition and multiplication with $\gamma(a) := \partial a$, or equiv. the gauge of any polynomial is its degree

4 Indication. Fact 3 is based on the properties of the degree of a polynomial and the Euclidean Division Theorem. Let $a, b \in \mathbb{R}[s]$ with $b \neq 0$; then $\exists!$ q, r in $\mathbb{R}[s]$ s.t.

$$a = bq + r \quad \text{with} \quad r = 0 \quad \text{or} \quad \partial r < \partial b.$$

Note that (a) if $\partial a < \partial b$, then $q = 0$ and $r = a$, and (b) by convention $\partial 0 := -\infty$.

5 Exercise. Show that the division of $a(s) := 4s^2 + 3s + 1$ by $b(s) := 2s + 1$ gives $q(s) = 2s + .5$ and $r(s) = .5$.

In control theory the following are also important Euclidean rings.

6 Fact [Hun.1] [Cal.1]. Consider $R(0)$ the ring of exp. stable transfer functions:

$$R(0) := \{f \in \mathbb{R}_p(s) : f \text{ is analytic in } \mathbb{C}_+\}.$$

Then $(R(0), +, \cdot, 0, 1)$ is a Euclidean ring under pointwise addition and pointwise multiplication where the gauge of any element $f \in R(0)$ is the number of zeros of f in $\mathbb{C}_+$ and at ∞ (the number of zeros of f at ∞ is the difference in degree between the denominator and the numerator of f).

7 Fact. Any field $(\mathbb{F}, +, \cdot, 0, 1)$ is a Euclidean ring with a gauge γ s.t. $\gamma(a)$ = nonzero constant, $\forall a \in \mathbb{F}\backslash\{0\}$.

8 Comment. An arbitrary Euclidean ring is not a field because every nonzero element has not necessarily a multiplicative inverse (for example, $\mathbb{R}[s]$).

9 Fact. $a \in \mathbb{R}[s]$ has an inverse in $\mathbb{R}[s]$, equiv. $a^{-1} = 1/a \in \mathbb{R}[s]$, iff $a(s) \equiv k$, where k is a nonzero constant.

10 Fact. $f \in \mathcal{R}(0)$ has an inverse in $\mathcal{R}(0)$, equiv. $f^{-1} = 1/f \in \mathcal{R}(0)$, iff $f(s) \neq 0 \ \forall s \in \mathbb{C}_+$ and at ∞.

Every engineer has encountered the vector spaces $\mathbb{R}^n$, $\mathbb{R}(s)^n$ over the fields $\mathbb{R}$, $\mathbb{R}(s)$ $\cdots$. The set of polynomial n-vectors denoted by $\mathbb{R}[s]^n$ is closed under addition and multiplication by elements of the ring $\mathbb{R}[s]$, which is not a field $\cdots$.

15 Definition. Let $\mathcal{R}$ be a commutative ring. A module M over $\mathcal{R}$, denoted by $(M, \mathcal{R}, +, \cdot, \theta_M)$, is a set M together with a commutative ring $\mathcal{R}$ such that

M1. $(M, +, \theta_M)$ is an additive commutative group, equiv.

$\exists$ a binary operation + called addition and given by

$$+ : M \times M \to M : (m_1, m_2) \mapsto m_1 + m_2$$

s.t.

(a) Addition is associative and commutative

(b) $\exists! \ \theta_M \in M$ s.t.

$$\forall m \in M, \ m + \theta_M = \theta_M + m = m$$

(c) $\forall m \in M \ \exists! \ -m \in M$ s.t.

$$m + (-m) = (-m) + m = \theta_M.$$

(Notation: $\forall m_1, m_2 \in M$ we write $m_1 - m_2$ to denote $m_1 + (-m_2)$.)

M2. The module is closed under multiplication $\cdot$ by scalars, equiv.

$$\cdot : \mathcal{R} \times M \to M : (r, m) \mapsto rm,$$

where $\forall m \in M$

$$1m = m, \quad 0m = \theta_M.$$

M3. Addition and multiplication by scalars are related by distributive laws, viz.,

$$\forall m \in M,\ \forall r_1, r_2 \in R,\ (r_1 + r_2)m = r_1 m + r_2 m;$$

$$\forall m_1, m_2 \in M,\ \forall r \in R,\ r(m_1 + m_2) = rm_1 + rm_2.$$ ∎

The definitions and facts below are also needed to explain $\mathbb{R}[s]^n$.

16 Definitions. Let $(M, R, +, \cdot, \theta_M)$ be a module.
(a) $S \subset M$ is a submodule if $(S, R, +, \cdot, \theta_M)$ is a module under the same operations as for M.
(b) A submodule generated by a subset $S \subset M$ is the intersection of all submodules containing S, or equiv. the smallest submodule containing S.

17 Definitions. Let M_1 and M_2 be two modules over the same ring R. The product module $M_1 \times M_2$ over R is the module $(M_1 \times M_2, R, +, \cdot, (\theta_{M_1}, \theta_{M_2}))$ using componentwise addition and multiplication by scalars.

18 Fact. $(\mathbb{R}[s]^n, \mathbb{R}[s], +, \cdot, \theta_n)$ is a module: it is the n-fold product module of $(\mathbb{R}[s], \mathbb{R}[s], +, \cdot, 0)$: a module under pointwise addition and multiplication.

20 Definitions. Let (M, R) denote a module M over the ring R. Let I be a finite index set. (a) We say that $(m_i)_{i \in I} \subset M$ is a linearly dependent family of (M, R) iff

$\exists$ scalars $r_i \in R$, $i \in I$, not all zero, such that

$$\sum_{i \in I} r_i m_i = \theta_M.$$

(b) We say that $(m_i)_{i \in I}$ is a basis of (M, R) iff (1) $(m_i)_{i \in I}$ is a linearly independent family and (2) the smallest submodule generated by $(m_i)_{i \in I}$ is M. (Note that not all R-modules have a basis because R is not a field [Sig.1].)
(c) A module M which has a basis is called a free module; the dimension of a free module M is by definition the cardinality of any basis of M.

21 Fact. $(\mathbb{R}[s]^n, \mathbb{R}[s], +, \cdot, \theta_n)$ is a free module of dimension n having the basis

$$e_i = (0, \cdots, 0, 1, 0, \cdots, 0)^T \in \mathbb{R}[s]^n \qquad i \in \underline{n}.$$
$$\uparrow$$
ith-component

22 Important Comment. A vector space V over the field $\mathbb{F}$, i.e., $(V, \mathbb{F}, +, \cdot, \theta_V)$, is similarly defined as a module $(M, R, +, \cdot, \theta_M)$, and has similar definitions for "subspace," "subspace generated by $S \subset V$," "product space," "linear dependence," "basis," and "dimension." A vector space always has a basis [Sig.1].

23 Exercise. Show that

(a) A vector space $(V, \mathbb{F})$ is a module (M, R);

(b) $(\mathbb{R}[s]^n, \mathbb{R}(s), +, \cdot, \theta_n)$ is not a module;

(c) $(R(0)^n, R(0), +, \cdot, \theta_n)$ is a free module of dimension n; is a subset of the n-dimensional vector space $(\mathbb{R}(s)^n, \mathbb{R}(s), +, \cdot, \theta_n)$; $(R(0)^n, R(0))$ and $(\mathbb{R}(s)^n, \mathbb{R}(s))$ have a common basis.

(d) $(\mathbb{R}[s]^2, \mathbb{R}[s])$ is a submodule of $(\mathbb{R}[s]^3, \mathbb{R}[s])$; is the submoaule generated by $S = \{(1, 0, 0)^T; (0, 1, 0)^T\}$.

(e) Show that $S := \{(s, 0), (0, 1)\}$ is not a basis for $(\mathbb{R}[s]^2, \mathbb{R}[s])$, but constitutes a linearly independent family. Determine the submodule generated by S.

Since the ring of polynomials $\mathbb{R}[s]$ is contained in the field of rational functions $\mathbb{R}(s)$, we have the following important fact.

26 Theorem. Consider the module $(\mathbb{R}[s]^n, \mathbb{R}[s], +, \cdot, \theta_n)$. Then,

(a)

27 $$(\mathbb{R}[s]^n, \mathbb{R}[s], +, \cdot, \theta_n) \subset (\mathbb{R}(s)^n, \mathbb{R}(s), +, \cdot, \theta_n)$$

We say that the $\mathbb{R}[s]$-module of polynomial n-vectors is embedded in the $\mathbb{R}(s)$-vector space of rational function n-vectors;

(b) with $(m_i)_{i \in I}$ a finite family of elements of the module $(\mathbb{R}[s]^n, \mathbb{R}[s])$

28 $(m_i)_{i \in I}$ is linearly dependent over $\mathbb{R}[s]$

$\Leftrightarrow$

29 $(m_i)_{i \in I}$ is linearly dependent over $\mathbb{R}(s)$

Comment. (b) means also that $(m_i)_{i \in I}$ is linearly independent (ℓ.i.) over $\mathbb{R}[s]$ iff this holds over the field $\mathbb{R}(s)$; consequently, the $\mathbb{R}[s]$-submodule generated by $(m_i)_{i \in I}$ inherits the dimension of the $\mathbb{R}(s)$-subspace generated by

$(m_i)_{i\in I}$: a question handled by classical vector space methods.

Proof of Theorem 26. (a) is obvious.

(b) $\Rightarrow$: is also obvious since $\mathbb{R}[s] \subset \mathbb{R}(s)$.

$\Leftarrow$: By assumption $\exists$ scalars $r_i \in \mathbb{R}(s)$, $i \in I$, not all zero s.t. $\sum_{i\in I} r_i m_i = \theta_r$. Let d be a least common denominator of the r_i's; then $dr_i \in \mathbb{R}[s]$, $\forall i \in I$, and the dr_i's are not all zero s.t. $\sum_{i\in I} dr_i m_i = \theta_n$. ∎

A polynomial matrix $M \in \mathbb{R}[s]^{m\times n}$ is the representation of a linear map sending the module $(\mathbb{R}[s]^n, \mathbb{R}[s])$ into the module $(\mathbb{R}[s]^m, \mathbb{R}[s])$. We say that M is an $m \times n$ matrix over the ring $\mathbb{R}[s]$. Note that matrices over a ring do not have the same properties as matrices over a field: in particular, invertibility, as shown presently.

33 Theorem. Consider $M \in \mathbb{R}[s]^{n\times n}$ as a linear map over $\mathbb{R}[s]$-modules, i.e., $M : (\mathbb{R}[s]^n, \mathbb{R}[s]) \to (\mathbb{R}[s]^n, \mathbb{R}[s]) : x \mapsto Mx$

U.t.c.

(a)

34 $\det M \neq 0 \Leftrightarrow$ the columns of M are linearly independent over $\mathbb{R}[s]$

$\Leftrightarrow Mx = \theta_r \Rightarrow x = \theta_n$

$\Leftrightarrow$ linear map M is injective (1 to 1).

(b)

35 $\det M = k$, a nonzero constant, $\Leftrightarrow M^{-1} \in \mathbb{R}[s]^{n\times n}$,

$\Leftrightarrow$ {linear map M is injective and surjective (onto).

36 Comments. (a) Condition (34) is necessary and sufficient for M to be invertible over the field $\mathbb{R}(s)$ but not over the ring $\mathbb{R}[s]$; note (35). Consider $M = \text{diag}[s + 1, s + 2] \in \mathbb{R}[s]^{2\times 2}$, then $M = \text{diag}[(s + 1)^{-1}, (s + 2)^{-1}] \in \mathbb{R}(s)^{2\times 2}$ but $\notin \mathbb{R}[s]^{2\times 2}$: the map $M : \mathbb{R}[s]^2 \to \mathbb{R}[s]^2$ is not surjective.

(b) Similar results hold for M^T, i.e., the transpose of M.

37 Definition. A polynomial matrix $M \in \mathbb{R}[s]^{n\times n}$ satisfying (34), resp. (35) is called nonsingular, resp. unimodular or invertible (nonsingular polynomial matrices are not necessarily invertible although their rational inverse exists!)

38 Proof of Theorem 33. (a) According to Theorem 26 one has: the columns of M are ℓ.i. over $\mathbb{R}[s]$ iff they are ℓ.i. over $\mathbb{R}(s)$. Now $M \in \mathbb{R}(s)^{n\times n}$: a matrix over a field. So the latter condition is equivalent to det $M \neq 0_{\mathbb{R}(s)} = 0_{\mathbb{R}[s]}$.
(b) According to Fact B.7, $M^{-1} \in \mathbb{R}[s]^{n\times n}$ iff $(\det M)^{-1} \in \mathbb{R}[s]$: the latter condition is equivalent to det $M \equiv k$ a nonzero constant by Fact 9. ∎

2.3. Polynomial Matrices

In this section we consider polynomial matrices $M \in \mathbb{R}[s]^{m\times n}$. All algebraic definitions and facts below hold for matrices $M \in \mathcal{R}^{m\times n}$, where $\mathcal{R}$ is a Euclidean ring modulo some small modifications. Proofs concerning the rank are specific for polynomial matrices.

2.3.1. Divisors, Coprimeness, Rank

We first discuss factoring a polynomial matrix. Avoid reading parallel definitions in a first reading.

1 Definitions. (a) Let A, B, $C \in \mathcal{E}(\mathbb{R}[s])$ with $A = BC$. Then C is said to be a right divisor (r.d.) of A and A is a left multiple (ℓ.m.) of C; similarly, B is a left divisor (ℓ.d.) of A and A is a right multiple (r.m.) of B.
(b) Similarly using matrices in $(\mathbb{R}[s])$, let $A = A_1R = LA_2$ and $B = B_1R = LB_2$. Then R (L) is said to be a common right divisor (c.r.d.) (common left divisor (c.ℓ.d.)) of A and B. If in addition R (L) is a ℓ.m. (r.m.) of every c.r.d. (c.ℓ.d.) of A and B, then R (L) is said to be a greatest common right divisor (g.c.r.d.) (greatest common left divisor (g.c.ℓ.d.)) of A and B.
(c) Two matrices A, $B \in \mathcal{E}(\mathbb{R}[s])$ with the same number of columns (rows) are said to be right-coprime (r.c.) (left-coprime (ℓ.c.)) iff they have a g.c.r.d. (g.c.ℓ.d.) which is unimodular.

2 Comment. A better terminology for divisor is factor: what is really meant is divisor without remainder.

3 Exercise. Show that if R is a g.c.r.d. of A and B as in Definition 1, then LR, with L a unimodular polynomial matrix, is also a g.c.r.d. (i.e., g.c.r.d.'s are not unique).

We next discuss the rank of a polynomial matrix.

5 Definitions. Let $M \in \mathbb{R}[s]^{m\times n}$ and let r be an integer s.t. $0 \leq r \leq \min(m,n)$.
(a) We say that M has normal (determinantal) rank r iff $\exists$ at least one $r \times r$ minor which is not the zero polynomial and $\forall s > r$ every $s \times s$ minor is the zero polynomial. We say also that M has rank r over $\mathbb{R}[s]$. This is denoted by

6 $$\text{rk } M = r.$$

(b) We say that M has local rank r at $s \in \mathbb{C}$ iff the matrix $M(s) \in \mathbb{C}^{m\times n}$ has rank r. This is denoted by

7 $$\text{rk}[M(s)] = r \quad \text{or} \quad \text{rk}M = r \text{ at } s.$$ ∎

8 Definition. For $M \in \mathbb{R}(s)^{m\times n}$ the normal rank is defined as rank over the field $\mathbb{R}(s)$. For any $s \in \mathbb{C}$ not a pole of M the local rank of M at s is the rank of $M(s) \in \mathbb{C}^{m\times n}$.

10 Theorem. Let $M \in \mathbb{R}[s]^{m\times n}$; then

6 $\text{rk}M = r$

$\Leftrightarrow$

11 r = maximum # of ℓ.i. columns of M in $(\mathbb{R}[s]^m, \mathbb{R}[s])$(=: normal column rank),

$\Leftrightarrow$

12 r = maximum # of ℓ.i. rows of M in $(\mathbb{R}[s]^n, \mathbb{R}[s])$(=: normal row rank).

13 Comment. By Theorem 10 for any polynomial matrix $M \in \mathbb{R}[s]^{m\times n}$ the normal determinantal rank = the normal column rank = the normal row rank: this common rank is called the normal rank of M.

14 Proof of Theorem 10. Observe that $M \in \mathbb{R}[s]^{m\times n} \subset \mathbb{R}(s)^{m\times n}$. Now for matrices over a field the assertion of the comment is true. The theorem follows therefore because (1) the determinantal rank over $\mathbb{R}[s]$ equals the determinantal rank over $\mathbb{R}(s)$ and (2) by Theorem 2.2.26 the maximum # of ℓ.i. columns (rows) over $\mathbb{R}[s]$ equals the maximum # of ℓ.i. columns (rows) over $\mathbb{R}(s)$. ∎

From the proof of Theorem 10, one has also

15 Theorem. Let $M \in \mathbb{R}[s]^{m \times n}$; then the rank of M over $\mathbb{R}[s]$ = the rank of M over $\mathbb{R}(s)$, i.e., M has the same normal rank as a polynomial and as a rational matrix.

16 Comment. Note that by Theorem 15 linear independence over the field $\mathbb{R}(s)$ may be used to check the normal row or column rank of M as a polynomial matrix.

Another consequence of Theorem 10 is

17 Exercise. Let $M \in \mathbb{R}[s]^{m \times n}$. Show that $\text{rk}M = \text{rk}M^T$.

The exercises below relate normal rank, invertibility and local rank.

18 Exercise. Let $M \in \mathbb{R}[s]^{m \times n}$. Show that

$$\text{rk}M = r \Leftrightarrow \text{rk}[M(s)] = r \text{ except for at most a finite number of points } s \in \mathbb{C}.$$

(Hint: Nonzero minors have at most a finite number of zeros.) ∎

20 Exercise. Let $M \in \mathbb{R}[s]^{n \times n}$. Show that

(a)

$$M \text{ is nonsingular} \Leftrightarrow \text{rk}M = n$$

$$\Leftrightarrow \text{rk}[M(s)] = n \text{ except for at most a finite number of points } s \in \mathbb{C}.$$

(b)

$$M \text{ is unimodular} \Leftrightarrow \text{rk}[M(s)] = n \ \forall s \in \mathbb{C}.$$ ∎

22 Note. If $M \in \mathbb{R}[s]^{m \times n}$ and $\text{rk}M = m$, ($\text{rk}M = n$), then M is said to have full row rank (full column rank) over $\mathbb{R}[s]$.

2.3.2. Elementary Operations on Polynomial Matrices

In this section we list elementary operations and their properties.

1 Let $M \in \mathbb{R}[s]^{m \times n}$. Elementary row operations (e.r.o.'s) on M are of three kinds:

(1) Interchange two rows $\rho_i \leftrightarrows \rho_j$.
(2) Multiply a row by a nonzero constant k (i.e., an invertible element of $\mathbb{R}[s]$): $\rho_i \leftarrow k\rho_i$.
(3) For $j \neq i$ add to row i another row j multiplied by $r \in \mathbb{R}[s]$: $\rho_i \leftarrow \rho_i + r\rho_j$.

Note that e.r.o.'s are equivalent to premultiplying M by left elementary matrices (ℓ.e.m.'s) L; these are obtained from the unit matrix by performing the desired e.r.o. upon it. The ℓ.e.m.'s corresponding to the e.r.o.'s above are listed below; any entry not explicitly shown is zero; dots indicate entries which are 1; broken lines indicate affected rows and columns.

$$(1):\ L = \begin{array}{c} \\ \\ \\ \\ i \\ \\ \\ \\ j \\ \\ \\ \\ \end{array} \left[\begin{array}{ccccccccccc} & & & i & & & & j & & & \\ 1 & & & & & & & & & & \\ & \ddots & & & & & & & & & \\ & & 1 & & & & & & & & \\ & & & 0 & & & & 1 & & & \\ & & & & 1 & & & & & & \\ & & & & & \ddots & & & & & \\ & & & & & & 1 & & & & \\ & & & 1 & & & & 0 & & & \\ & & & & & & & & 1 & & \\ & & & & & & & & & \ddots & \\ & & & & & & & & & & 1 \end{array}\right] \in \mathbb{R}[s]^{m\times m},$$

$$(2):\ L = \begin{array}{c} \\ \\ \\ \\ i \\ \\ \\ \end{array} \left[\begin{array}{cccccc} & & & i & & \\ 1 & & & & & \\ & \ddots & & & & \\ & & 1 & & & \\ & & & k & & \\ & & & & 1 & \\ & & & & & 1 \end{array}\right] \in \mathbb{R}[s]^{m\times m},$$

$$(3):\ L = \begin{array}{c} \\ \\ \\ i \\ \\ j \\ \\ \\ \end{array} \left[\begin{array}{ccccccc} & & i & & j & & \\ 1 & & & & & & \\ & \ddots & & & & & \\ & & 1 & & r & & \\ & & & \ddots & & & \\ & & & & 1 & & \\ & & & & & \ddots & \\ & & & & & & 1 \end{array}\right] \in \mathbb{R}[s]^{m\times m}.$$

2 Elementary column operations (e.c.o.'s) are similarly defined: replace "row" by "column" ($\rho_i \leftarrow \gamma_i$; $\rho_j \leftarrow \gamma_j$). E.c.o.'s are equivalent to

postmultiplying M by right elementary matrices (r.e.m.'s) R; these are obtained from the unit matrix by performing the desired e.c.o. upon it.

3 Exercise. Write down the three e.c.o.'s and their corresponding r.e.m.'s $R \in \mathbb{R}[s]^{n\times n}$.

4 Definitions. (a) A square polynomial matrix is said to be an elementary matrix (e.m.) iff it is a ℓ.e.m. or a r.e.m..
(b) An operation on a polynomial matrix is said to be an elementary operation (e.o.) iff it is an e.r.o. or an e.c.o..

Elementary operations are used to reduce polynomial matrices to forms that display wanted information; see Sec. 2.3.4 on standard forms. The following are important properties.

7 Exercise. Show that each e.o. is invertible, hence each e.m. is unimodular (equiv. invertible).
(Hint: $\rho_j \rightleftarrows \rho_i$; $\rho_i \leftarrow k^{-1}\rho_i$; ····; det L = k a nonzero constant ····.)

8 Theorem [Rank invariances]. Each elementary operation on a polynomial matrix leaves the normal rank unchanged.

Equivalently, let $M \in \mathbb{R}[s]^{m\times n}$ be a polynomial matrix and let $L \in \mathbb{R}[s]^{m\times m}$ and $R \in \mathbb{R}[s]^{n\times n}$ be an arbitrary left-, resp. right-, elementary matrix; then

$$\mathrm{rk}M = \mathrm{rk}LM, \quad \mathrm{rk}M = \mathrm{rk}MR.$$

9 Comment. This property is a key tool for the reduction to standard forms.

10 Proof. Note (a) $M \in \mathbb{R}(s)^{m\times n}$: a matrix over a field; (b) the e.o.'s listed above are also e.o.'s for matrices over the field $\mathbb{R}(s)$; (c) the latter operations leave the rank over $\mathbb{R}(s)$ unchanged; (d) by Theorem 2.3.2.15 the rank over $\mathbb{R}(s)$ is the rank over $\mathbb{R}[s]$. Hence the theorem. ∎

The last part of this section catches e.o.'s into the compact notion of equivalence. A forward reference is necessary.

13 Theorem. Let $M \in \mathbb{R}[s]^{m\times m}$.

M is unimodular

$\Leftrightarrow$

M is a finite product of elementary matrices.

Proof. $\Leftarrow$: follows by Exercise 7.

$\Rightarrow$: If M is unimodular, then it can be reduced by e.o.'s to a Smith form which is the unit matrix: see Exercise 2.3.4.30 below. ∎

14 Definitions. Let A and B $\in \mathbb{R}[s]^{m\times n}$; then A and B are said to be equivalent, left equivalent, right equivalent iff there exist unimodular matrices $L \in \mathbb{R}[s]^{m\times m}$ and $R \in \mathbb{R}[s]^{n\times n}$ s.t. resp. A = LBR, A = LB, A = BR. These relations are denoted by resp. $A \sim B$, $A \overset{\ell}{\sim} B$, $A \overset{r}{\sim} B$.

15 Exercise. Show that the relations defined in Definitions 14 are equivalence relations.
(Hint: $A \sim A$; $A \sim B \Rightarrow B \sim A$; $A \sim B$, $B \sim C \Rightarrow A \sim C$; $\cdots$.)

16 Exercise. With matrices A and B as in Definitions 14, show that A and B are equivalent, left equivalent, right equivalent iff A and B can be obtained from B and A by resp. e.o.'s, e.r.o.'s, e.c.o.'s.
(Hint: Use Theorem 13.)

17 Exercise. Show that the relations of equivalence, left equivalence, and right equivalence do not change the rank.
(Hint: Use Exercise 16 and Theorem 8.)

2.3.3. Elementary Operations and Differential Equations

Consider the differential equation

1 $$D(p)\xi(t) = \theta_\nu \quad \forall t \geq 0$$

where p = d/dt is the differential operator, $D(\cdot) \in \mathbb{R}[p]^{\nu\times\nu}$ is a nonsingular polynomial matrix in p. For example,

2 $$D(p) = D_2p^2 + D_1p + D_0 \qquad D_i \in \mathbb{R}^{\nu\times\nu} \text{ for } i = 0 \sim 2.$$

Note that (1) is a differential equation with constant coefficients; for such an equation it is well known that any solution $\xi(\cdot)$ with values in $\mathbb{R}^\nu$ is a sum of exponential polynomials in t of the form

$$\sum_{k=1}^{m} (a_k t^{k-1})\exp(\lambda t), \quad \forall t > 0-,$$

where $a_k \in \mathbb{C}^{\nu}$ $\forall k \in \underline{m}$, $\lambda \in \mathbb{C}$ and $\det D(\lambda) = 0$ with the restriction that iff $\lambda \in \mathbb{C}\setminus\mathbb{R}$, then the solution has a complex conjugate companion term given by

$$\sum_{k=1}^{m} (\bar{a}_k t^{k-1})\exp(\bar{\lambda} t), \quad \forall t > 0-.$$

As a consequence, solutions of (1) are <u>infinitely differentiable on $(0-,\infty)$</u>: in particular, $t \to \xi(t)$ and all its derivatives are continuous at $t = 0$.

The reason we emphasize the behavior of $\xi(\cdot)$ at $t = 0$ is that when the input $u(\cdot)$ will be introduced, the solution of $D(p)\xi(t) = N_{\ell}(p)u(t)$, may be discontinuous at $t = 0$: hence it will be important then to distinguish between $\xi(0-)$ and $\xi(0+)$.

Hence the following definition.

3 <u>Definition</u>. A <u>solution</u> of (1) is a function $\xi(\cdot)$: $(0-,\infty) \to \mathbb{R}^{\nu}$ s.t. $D(p)\xi(t) = \theta_{\nu}$ $\forall t \geq 0$. The set of solutions of (1) is denoted by X.

4 <u>Fact</u>. The set X of solutions of (1) is an <u>$\mathbb{R}$-vector space</u>, equiv. if $\xi_1(\cdot)$, $\xi_2(\cdot) \in X$ and a_1, $a_2 \in \mathbb{R}$, then $a_1\xi_1(\cdot) + a_2\xi(\cdot) \in X$.

5 <u>Exercise</u>. Prove Fact 4. (Hint: Use (2) and observe that real matrix- and differential operators are linear.)

The consequences of e.o.'s over $\mathbb{R}[p]$ on polynomial matrix $D(p)$ in (1) are now explained. We start with e.r.o.'s.

6 <u>Theorem</u>. Let $D(\cdot) \in \mathbb{R}[p]^{\nu\times\nu}$ be nonsingular and let $L(\cdot) \in \mathbb{R}[p]^{\nu\times\nu}$ be <u>unimodular</u> with

7 $$\bar{D}(p) = L(p)D(p).$$

Consider the differential equations

8 $$D(p)\xi(t) = \theta_{\nu} \quad t \geq 0; \qquad \bar{D}(p)\bar{\xi}(t) = \theta_{\nu} \quad t \geq 0$$

with solution spaces X resp. $\bar{X}$.

U.t.c.

9 $$X = \bar{X}.$$

10 Comment. Performing e.r.o.'s on $D(p)$ does not affect the solution space of $D(p)\xi(t) = \theta, t \geq 0$.

Proof of Theorem 6. Since L is unimodular both $L(\cdot)$ and $L(\cdot)^{-1}$ are polynomial matrices, whence $\forall t \geq 0$

$$D(p)\xi(t) = \theta \;\Rightarrow\; L(p)D(p)\xi(t) = \bar{D}(p)\xi(t) = \theta;$$

$$\bar{D}(p)\bar{\xi}(t) = \theta \;\Rightarrow\; L(p)^{-1}\bar{D}(p)\bar{\xi}(t) = D(p)\bar{\xi}(t) = \theta.$$ ∎

18 Theorem. Let $D(\cdot) \in \mathbb{R}[p]^{\nu\times\nu}$ be nonsingular and $R(\cdot) \in \mathbb{R}[p]^{\nu\times\nu}$ be unimodular with

19 $$\bar{D}(p) = D(p)R(p).$$

Consider the differential equations

20 $$D(p)\xi(t) = \theta_\nu \quad t \geq 0; \qquad \bar{D}(p)\bar{\xi}(t) = \theta_\nu \quad t \geq 0$$

with solution spaces X resp. $\bar{X}$.
U.t.c. the map

21 $$T : \bar{\xi}(\cdot) \in \bar{X} \mapsto R(p)\bar{\xi}(\cdot) =: \xi(\cdot) \in X$$

is a linear bijection (i.e., an isomorphism) from $\bar{X}$ onto X.

22 Comments. E.c.o.'s on $D(p)$ result in a "change of variables" for the solutions of $D(p)\xi(t) = \theta$, $t \geq 0$. ∎

23 Proof of Theorem 18. Note that since $R(\cdot)$ is unimodular, $R(\cdot)$ and $R(\cdot)^{-1}$ are polynomial matrices. Hence

(a) $\forall t \geq 0$ with $\xi(\cdot) := R(p)\bar{\xi}(\cdot)$, where $\bar{\xi}(\cdot) \in \bar{X}$,

$$D(p)\xi(t) = D(p)R(p)\bar{\xi}(t) = \bar{D}(p)\bar{\xi}(t) = \theta.$$

Hence T defined by (21) maps $\bar{X}$ into X.

(b) $\forall t \geq 0$ with $\bar{\xi}(\cdot) = R(p)^{-1}\xi(\cdot)$, where $\xi(\cdot) \in X$,
$\bar{D}(p)\bar{\xi}(t) = \bar{D}(p)R(p)^{-1}\xi(t) = D(p)\xi(t) = \theta$.

Hence T defined by (21) is invertible.

The linearity of map (21) is left as an exercise. ∎

The last theorem of this section concerns itself with the effects of arbitrary e.o.'s, i.e., equivalence on $D(\cdot)$.

26 Theorem. Let $D(\cdot) \in \mathbb{R}[p]^{\nu\times\nu}$ be nonsingular and let $L(\cdot) \in \mathbb{R}[p]^{\nu\times\nu}$ and $R(\cdot) \in \mathbb{R}[p]^{\nu\times\nu}$ be unimodular matrices with

27 $$\tilde{D}(p) = L(p)D(p)R(p).$$

Consider the differential equations

28 $$D(p)\xi(t) = \theta_\nu, t \geq 0; \quad \tilde{D}(p)\bar{\xi}(t) = \theta_\nu, t \geq 0$$

with solution spaces X resp. $\bar{X}$.
U.t.c. the map

29 $$T : \bar{\xi}(\cdot) \in \bar{X} \mapsto R(p)\bar{\xi}(\cdot) =: \xi(\cdot) \in X$$

is a linear bijection (isomorphism) of $\bar{X}$ onto X. ∎

30 Exercise. Prove Theorem 26.
(Hint: Use Theorems 6 and 18.)

31 Comment. (a) In other words, $\bar{\xi}(\cdot)$ is a solution of (28^2) iff $\xi(\cdot)$ defined by (29) is a solution of (28^1).
(b) Isomorphic vector spaces have the same dimension; an isomorphism maps a basis onto a basis; etc.

2.3.4. Standard Forms: Hermite and Smith Forms

Every polynomial matrix $M \in \mathbb{R}[s]^{m\times n}$ can by e.o.'s be brought into a standard form: the most important are the Hermite form (upper or lower triangular) and the Smith form (diagonal).

nonzero entries below the degree of the leading entry. ∎

The Hermite row form is obtained by the following algorithm.

5.L. Algorithm [Reduction to Hermite row form]. For $i = 1, 2, \cdots$:

Step 1. Search for γ_{p_i}, the first column from the left that is nonzero below ρ_{i-1}.
If such column does not exist, or $i = m+1$, or $p_{i-1} = n$, STOP.

Step 2. Choose among the nonzero entries of γ_{p_i} below ρ_{i-1} an entry of smallest degree and by row permutation place it in position (i,p_i).

Step 3. Multiply ρ_i by a nonzero constant to make the entry in position (i,p_i) monic.

Step 4. Use the Euclid algorithm and addition of suitable polynomial multiples of ρ_i to reduce the nonzero entries in γ_{p_i} to their remainders after division by entry (i,p_i).

Step 5. If the remainders in γ_{p_i} below ρ_i are all zero, go to step 6. Else, repeat steps 2 to 4 until the remainders are all zero.

Step 6. If $i = 1$, skip step 6. Else, use the Euclid algorithm and addition of suitable polynomial multiples of ρ_i to reduce nonzero entries in γ_{p_i} above ρ_i to their remainders after division by entry (i,p_i). ∎

6 Exercise. Reduce to Hermite row form

$$M(s) = \begin{bmatrix} 0 & 0 & s^2 \\ s+1 & 0 & s^2 \\ s+2 & 0 & -s^2 \\ s+1 & 0 & -s^2 \end{bmatrix}.$$

7 Proof of Theorem 1.L. The proof is by construction and consists in justifying Algorithm 5.L.

A. Concerning step 5, note that the degrees of the nonzero remainders of the nonzero entries of γ_{ρ_i} below row i are smaller than the degree of entry (i,ρ_i). Hence by repeating steps 2 to 4 these degrees are steadily decreasing and zero remainders are ultimately obtained: step 6 will be reached after a finite number of sequences consisting of steps 2 to 4.

B. Since matrix M has a finite number of entries, the algorithm will stop in step 1 for a finite value of i =: r + 1 and the matrix will have a staircase form with exactly r stairs occurring in ρ_i, $\forall i \in \underline{r}$, as in (2).

C. Over the field $\mathbb{R}(s)$ the rank of H in (2) is clearly r. Since by Theorem 2.3.1.15 this is also the rank of H over $\mathbb{R}[s]$ and by Theorem 2.3.2.8 e.o.'s do not change the rank over $\mathbb{R}[s]$, we have

$$r = \text{rk}H = \text{rk}M.$$

Hence (3) holds.

D. It follows finally by construction that all properties of H listed under (a)-(c) of the theorem statement hold. ∎

8 Exercise. Read [e.g., Nob.1,pp.82-83] and observe that Algorithm 5.L is the Gauss-Jordan algorithm for obtaining the row-echelon form of $M \in \mathbb{R}^{m\times n}$ modulo modifications involving the use of the Euclid algorithm to reduce the degree of a polynomial and eventually making a polynomial zero.

9.L. Corollary. [Hermite row form of a full column rank matrix]. Let in Theorem 1.L $M \in \mathbb{R}[s]^{m\times n}$ have full column rank over $\mathbb{R}[s]$, equiv. rkM = n, whence $m \geq n$. Then there exists a unimodular matrix $L \in \mathbb{R}[s]^{m\times m}$ (obtained by e.r.o.'s) such that

$$LM = H = \overset{n}{\begin{bmatrix} R \\ \\ 0 \end{bmatrix}} \begin{matrix} n \\ \\ m-n \end{matrix}$$

with $R \in \mathbb{R}[s]^{n\times n}$ upper triangular and nonsingular; i.e., H, the Hermite row form of M, is upper triangular. Moreover, H is as in the statement of Theorem 1.L with

$$p_i = i \quad \forall i \in \underline{n}.$$

Proof. By the proof of Theorem 1.L, r = rkH is the number of stairs of the Hermite row form H given by (2). By assumption rkM = n, whence by (3) r = rkH = rkM = n. Hence matrix $H \in \mathbb{R}[s]^{m\times n}$ given by (2) must have exactly n stairs. The corollary follows. ∎

12 Convention. Up to now we obtained the Hermite row form of a polynomial matrix described by Theorem 1.L [Hermite row form], Algorithm 5.L [Reduction to Hermite row form] and Corollary 9.L [Hermite row form of a full column rank matrix]. Note that L stands for left operations (e.r.o.'s). Using e.c.o.'s instead of e.r.o.'s we have also Theorem 1.R [Hermite column form], Algorithm 5.R [Reduction to Hermite column form], and Corollary 9.R [Hermite column form of a full row rank matrix], where R stands for right operations (e.c.o.'s). This convention is used below. For reasons of completeness we give Theorem 1.R in detail.

1.R. Theorem [Hermite column form]. Let $M \in \mathbb{R}[s]^{m\times n}$. Then there exists a unimodular matrix $R \in \mathbb{R}[s]^{m\times n}$ (obtained by e.c.o.'s) such that

13 $MR = H =$

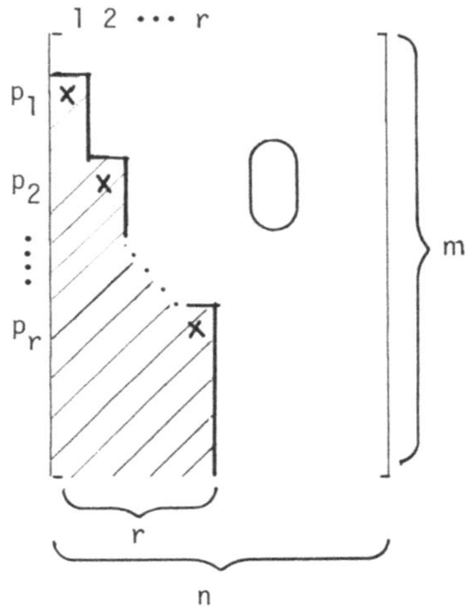

where H is called Hermite column form of M and has the following properties:

There exists an integer r, $0 \le r \le \min(m, n)$ with

14 $$\text{rk}H = \text{rk}M = r$$

s.t.

(a) $\forall j \in \underline{r}$, γ_j has a leading nonzero monic polynomial $h_{p_j j}$ called leading entry s.t.

$$1 \le p_1 < p_2 < \cdots < p_r \le m.$$

(b) $\forall j \in \underline{r}$,

$$\text{if } h_{p_j j} = 1, \text{ then } h_{p_j i} = 0 \ \ \forall i < j,$$

$$\text{if } h_{p_j j} \neq 1, \text{ then } \partial[h_{p_j i}] < \partial[h_{p_j j}] \ \ \forall i < j \ \text{ s.t. } \ h_{p_j i} \neq 0.$$

(c) $\forall\ \rho_i$ s.t. $i < p_1$, ρ_i is zero;

$\forall\ \rho_i$ s.t. $p_j \le i < p_{j+1}$ with $j \in \underline{r-1}$, then the last $m - j$ elements of ρ_i are zero;

$\forall\ \rho_i$ s.t. $i \ge p_r$, the last $m - r$ elements of ρ_i are zero. ∎

15 Exercise. Write the statements of Algorithm 5.R [Reduction to Hermite column form] and Corollary 9.R [Hermite column form of a full row rank matrix]. (Hint: Use the statements of Algorithm 5.L and Corollary 9.L; replace e.r.o.'s by e.c.o.'s, "row" by "column," ···.)

Instead of using only e.r.o.'s (Hermite row form) or only e.c.o.'s (Hermite column form) we use now a combination of e.r.o.'s and e.c.o.'s to obtain a quasi-diagonal standard form.

18 Theorem [Smith form]. Let $M \in \mathbb{R}[s]^{m\times n}$. Then there exist unimodular matrices $L \in \mathbb{R}[s]^{m\times m}$ (obtained by e.r.o.'s) and $R \in \mathbb{R}[s]^{n\times n}$ (obtained by e.c.o.'s) such that

19 $$LMR = S = \left[\begin{array}{ccccccc|c} \lambda_1 & & & & & & & \\ & \ddots & & & O & & & O \\ & & \lambda_i & & & & & \\ & & & \ddots & & & & \\ & O & & & & \lambda_r & & \\ \hline & & & O & & & & O \end{array}\right] \begin{array}{l} \} r \\ \\ \\ \\ \\ \} m-r \end{array} \in \mathbb{R}[s]^{m\times n},$$
(column blocks: r, $n-r$)

where S is called the Smith form of M and

(a)

20 $$r = \text{rk}M = \text{rk}S \le \min(m, n).$$

(b) The polynomials $\lambda_i \in \mathbb{R}[s]$, $i \in \underline{r}$, called invariant polynomials of M, are monic, uniquely defined by M, and satisfy the division property

21 $$\lambda_i \mid \lambda_{i+1} \qquad \forall i \in \underline{r-1}.$$

Moreover, consider the polynomials $\Delta_i \in \mathbb{R}[s]$, $i = 0 \sim r$, given by

22 $\Delta_0 \equiv 1$, $\Delta_i :=$ the monic g.c.d. of all $i \times i$ minors of M.

These polynomials are called the <u>determinantal divisors</u> of M. They are related to the invariant polynomials of M by

23 $$\lambda_i = \Delta_i / \Delta_{i-1} \qquad i \in \underline{r}.$$ ∎

24 <u>Comment</u>. The top left block of the Smith form (19) is diagonal with unique invariant polynomials λ_i as diagonal entries: the Smith form is a <u>unique</u> standard form under equivalence.

25 <u>Proof of Theorem 18</u>. The proof is by construction and is carried out in four steps, A, B, C, and D.

26 A. <u>Algorithm</u> [Reduction to Smith form]

(1) By a succession of e.o.'s reduce matrix M into the form

27 $$\begin{array}{c} \\ 1\left\{\right. \\ \\ \\ \\ \end{array}\overset{\overbrace{\ \ }^{1}}{\left[\begin{array}{c|ccc} \lambda_1 & 0 & \cdots & 0 \\ \hline 0 & & & \\ \vdots & & M_1 & \\ 0 & & & \end{array}\right]}$$

where λ_1 is monic and divides every element of M_1. This is done as follows:

<u>Step 1</u>. By a row and/or column permutation place a lowest degree nonzero entry in position (1, 1).

<u>Step 2</u>. If all entries of γ_1 and ρ_1 except for entry (1, 1) are zero, go to step 4.

<u>Step 3</u>. Find the remainders after division by entry (1, 1) of all entries of γ_1 and ρ_1 except for entry (1, 1).

<u>If</u> all these remainders are zero, then by the addition of suitable multiples of ρ_1 and γ_1 reduce the corresponding nonzero entries to zero and

go to step 4.

Else consider a nonzero remainder of lowest degree and its corresponding nonzero entry. By the addition of a suitable multiple of ρ_1 or γ_1 replace that entry by its remainder and by a row or column permutation bring that remainder in position (1, 1). Next go to step 2.

Comment: The outcome of steps 2 and 3 is to reduce to zero every entry of γ_1 and ρ_1 except for entry (1, 1); note that the degree of entry (1, 1) steadily decreases when cycling occurs.

Step 4. *If* entry (1, 1) divides every entry of the matrix, STOP. *Else* there exists among the entries not in position (1, 1) a nonzero entry, say in γ_k, not divisible without remainder by entry (1, 1) and of lowest degree. Then add γ_k to γ_1 and repeat step 3. Keep on repeating until entry (1, 1) divides every entry of the matrix.

Comment: The degree of entry (1, 1) continues to decrease···. ∎

(2) Now operate similarly on M_1 in (27) to obtain

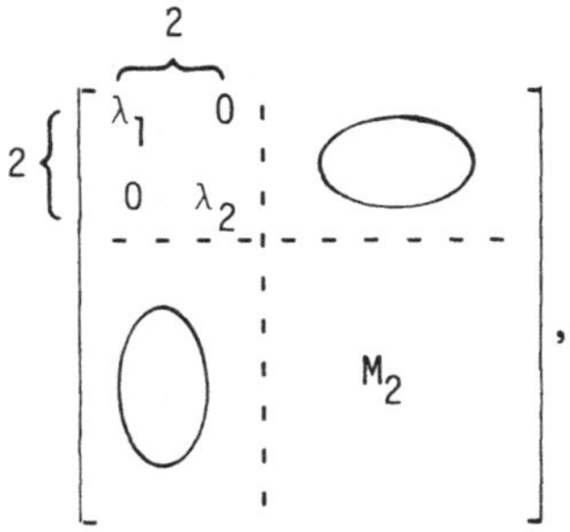

where λ_2 is monic and divides every element of M_2

⋮

(i) Operate similarly on M_{i-1} to obtain

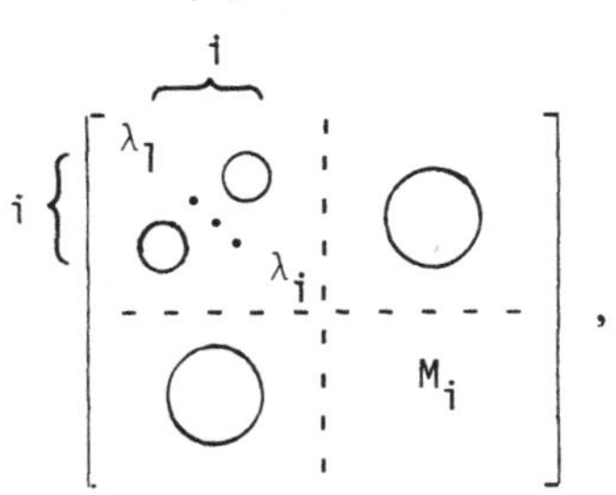

where λ_i is monic and divides every entry of M_i.

⋮

Do this until $i = \min(m, n)$ or $M_i = 0$, then STOP

End of Algo.

B. In the Smith Form Algorithm the sequence steps 1-4 is a finite procedure because the degree of the corner entry is steadily decreased. Moreover, the algorithm will stop when it runs out of submatrices M_i. Then, set $r :=$ the last value of i and the matrix will have form (19) with property (20).

C. In the Smith Form Algorithm, $\forall i \in \underline{r-1}$ there exist unimodular matrices L_i, R_i (obtained by e.o.'s) such that

$$L_i M_i R_i = \left[\begin{array}{c|ccc} \lambda_{i+1} & 0 & \cdots & 0 \\ \hline 0 & & & \\ \cdot & & & \\ \cdot & & M_{i+1} & \\ \cdot & & & \\ 0 & & & \end{array}\right]$$

and λ_i divides <u>every</u> entry of M_i. Obviously, λ_{i+1} is an $\mathbb{R}[s]$-linear combination of entries of M_i. Hence $\lambda_i | \lambda_{i+1}$: (21) holds.

D. Let

$\Delta_i^S :=$ the monic g.c.d. of all $i \times i$ minors of S,

$\Delta_i^M :=$ the monic g.c.d. of all $i \times i$ minors of M;

then by the equivalence relation (19)

$$LMR = S \quad \text{and} \quad M = L^{-1} S R^{-1},$$

where L, L^{-1}, R, and R^{-1} are polynomial square matrices. By the Cauchy-Binet formula [e.g., Kai. 1, pp. 649-650] it follows that every $i \times i$ minor of S is an $\mathbb{R}[s]$-linear combination of $i \times i$ minors of M and vice versa (exercise). Hence

$$\Delta_i^M | \Delta_i^S \quad \text{and} \quad \Delta_i^S | \Delta_i^M,$$

s.t.

$$\Delta_i^M = \Delta_i^S =: \Delta_i$$

or equiv. the monic g.c.d. of all $i \times i$ minors is invariant under equivalence. Now from the form of S in (19) and division property (21) $\Delta_i = \prod_{j=1}^{i} \lambda_j$ for $i \in \underline{r}$; hence, with $\Delta_0 \equiv 1$, (23) is established. Since the Δ_i are unique, by (23) the λ_i must also be unique and we are done. ∎

29 Exercise. Find the Smith form of

$$M(s) = \begin{bmatrix} s+1 & 0 & s(s+1) \\ 0 & s+2 & s(s+2) \end{bmatrix}$$

30 Exercise. Show that the Smith form of a unimodular matrix is the unit matrix. (This is used in the proof of Theorem 2.3.2.13.)

From the proof of Theorem 19 and from Exercise 2.3.2.16, we have also

31 Corollary [Characterization of equivalence]. Two polynomial matrices M and $N \in \mathbb{R}[s]^{m\times n}$ are equivalent, equiv. can be obtained from each other by e.o.'s, if and only if one the following conditions hold:

(a) M and N have the same Smith form;

(b) M and N have the same invariant polynomials;

(c) M and N have the same determinantal divisors. ∎

2.3.5. The Solution Space of $D(p)\xi(t) = \theta_\nu$, $t \geq 0$

In this section we continue the description of the $\mathbb{R}$-linear space X of solutions $\xi(\cdot) : (0-, \infty) \to \mathbb{R}^\nu$ of the differential equation

1 $$D(p)\xi(t) = \theta_\nu \quad t \geq 0,$$

where $D(\cdot) \in \mathbb{R}[p]^{\nu\times\nu}$ is nonsingular and $p = d/dt$ is the differential operator: see Sec. 2.3.3.

A first result involves the dimension of the solution space X.

2 Theorem [Dimension of the solution space]. Let $D(\cdot) \in \mathbb{R}[p]^{\nu\times\nu}$ be nonsingular and consider the $\mathbb{R}$-linear space X of solutions $\xi(\cdot) : (0-, \infty) \to \mathbb{R}^\nu$ of the differential equation

1 $$D(p)\xi(t) = \theta_\nu \quad t \geq 0$$

U.t.c.

3 $$\dim X = \partial[\det D]. \qquad ∎$$

4 Comment. (3) means that any solution of (1) is an $\mathbb{R}$-linear combination of a basis consisting of $n := \partial[\det D]$ linearly independent solutions: once a basis of X is known then all solutions are known.

5 Proof. From Theorem 2.3.3.26 and Comment 2.3.3.31, without loss of generality, we may assume that $D(\cdot)$ is in the Smith form of Theorem 2.3.4.18, i.e.,

6 $$D(p) = S(p) = \operatorname{diag}[\lambda_i(p)]_{i=1}^{\nu},$$

where the $\lambda_i(\cdot) \in \mathbb{R}[p]$ are the invariant polynomials of $D(\cdot)$. As a consequence with $\xi(\cdot) = (\xi_i(\cdot))_{i=1}^{\nu}$, $\xi_i(\cdot) : (0-, \infty) \to \mathbb{R}$, $\forall i \in \underline{\nu}$, the vector differential equation (1) decouples into ν scalar differential equations $\lambda_i(p)\xi_i(t) = 0$, $t \geq 0$ and the solution space X is the direct sum of component spaces X_i, $i \in \underline{\nu}$. More precisely,

7 $$X = \bigoplus_{i=1}^{\nu} X_i,$$

where

$$\begin{aligned} X_i = \{\xi(\cdot) : (0-, \infty) \to \mathbb{R}^{\nu};\ & \xi(\cdot) = (0, \cdots, 0, \underset{\substack{\uparrow \\ i\text{th component}}}{\xi_i(\cdot)}, 0, \cdots, 0)^T, \\ & \lambda_i(p)\xi_i(t) = 0 \quad \forall t \geq 0\} \end{aligned}$$

Now by the theory of scalar differential equations

8 $$\dim X_i = \partial[\lambda_i(\cdot)].$$

(A basis for X_i is generated by $\partial[\lambda_i(\cdot)]$ linearly independent solutions of $\lambda_i(p)\xi_i(t) = 0$, $t \geq 0$.) Hence by (6)-(8)

$$\dim X = \sum_{i=1}^{\nu} \dim X_i = \sum_{i=1}^{\nu} \partial[\lambda_i] = \partial[\det D]. \qquad ■$$

From the structure of vector spaces and Theorem 2 we have immediately

11 Theorem. Let $D(\cdot) \in \mathbb{R}[p]^{\nu\times\nu}$ be nonsingular and consider the differential equation

$$1 \qquad D(p)\xi(t) = \theta_\nu \qquad t \geq 0$$

with solution space X.

Let

$$12 \qquad n := \partial[\det D]$$

denote the dimension of X and let $(\psi_k(\cdot))_{k=1}^{n}$ denote any basis of X generating a basis matrix $\Psi(\cdot) : (0-, \infty) \to \mathbb{R}^{\nu\times n}$ given by

$$13 \qquad \Psi(t) = \left[\ \psi_1(t) \,\vdots\, \psi_2(t) \,\vdots\, \cdots \,\vdots\, \psi_n(t)\ \right], \text{ for } t > 0-.$$

Let $x(0) = (x_k(0))_{k=1}^{n}$ be any n-vector of $\mathbb{R}^n$

U.t.c.

(a) $\forall\ \xi(\cdot) \in X$, or equiv. $\forall$ solution of (1), there is a unique $x(0) \in \mathbb{R}^n$ s.t.

$$14 \qquad \xi(t) = \Psi(t)x(0), \qquad \forall t > 0-,$$

(b) the map

$$15 \qquad x(0) \in \mathbb{R}^n \mapsto \xi(\cdot) = \Psi(\cdot)x(0) \in X$$

is an $\mathbb{R}$-linear bijection (isomorphism) from $\mathbb{R}^n$ onto X. ■

Theorem 11 justifies the following definition.

16 Definition. We call state at $t = 0$ of differential equation (1) the n-vector $x(0) \in \mathbb{R}^n$ defined by equation (14) through the choice of a basis $(\psi_k(\cdot))_{k=1}^{n}$ for the solution space X.

17 Comments. (a) For every choice of basis the definition of the state at $t = 0$ is justified because of map (15): the knowledge of $x(0)$ completely determines the trajectory $\xi(\cdot)$; conversely to every trajectory corresponds a unique state $x(0)$ at $t = 0$.

(b) The definition of the state at $t = 0$ depends on the choice of a basis for the solution space X of (1): below we shall normalize this definition by finding a state at $t = 0$ related to initial data of $\xi(\cdot)$ and its derivatives: see Algorithm 36.

(c) For every choice of basis the dimension of the state space is

12 $$n := \partial[\det D].$$

(d) $x(0)$ is zero means that $\xi(\cdot)$ and all its derivatives are identically zero.
(e) For two arbitrary bases $(\psi_k(\cdot))_{k=1}^{n}$ and $(\bar{\psi}_k(\cdot))_{k=1}^{n}$ of the solution space X of (1), the relation between the basis matrices $\Psi(\cdot)$ and $\bar{\Psi}(\cdot)$ (defined as in (13)) is given by

18 $$\Psi(\cdot) = \bar{\Psi}(\cdot)P,$$

where $P \in \mathbb{R}^{n\times n}$ is nonsingular. As a consequence the states $x(0) \in \mathbb{R}^n$ and $\bar{x}(0) \in \mathbb{R}^n$ defined by

19 $$\xi(t) = \Psi(t)x(0) = \bar{\Psi}(t)\bar{x}(0), \quad \text{for } t > 0-,$$

are related by

20 $$\bar{x}(0) = P\, x(0),$$

where $P \in \mathbb{R}^{n\times n}$ is nonsingular.

(f) A convenient method to obtain a basis $(\psi_k(\cdot))_{k=1}^{n}$ of the solution space X of (1) is to reduce first $D(p)$ to the upper-triangular Hermite row form by Algorithm 2.3.4.5.L (leaving each solution $\xi(\cdot)$ and the vector space X unchanged; see Theorem 2.3.3.6), and then use Algorithm 24 below.

24 Algorithm [Construction of a basis of solutions]
Data: $D = [d_{ij}]_{i,j\in\underline{\nu}} \in \mathbb{R}[p]^{\nu\times\nu}$ is nonsingular and in upper-triangular Hermite row form as described by Theorem 2.3.4.1.L and Corollary 2.3.4.9.L.

We consider a differential equation

1 $$D(p)\xi(t) = \theta_\nu \quad t \geq 0,$$

or equiv. a system of differential equations

25 $$\sum_{j=i}^{\nu} d_{ij}(p)\xi_j(t) = 0 \quad t \geq 0 \quad \text{for } i \in \underline{\nu}$$

with solution space X. Define

26 $$n_i := \partial[d_{ii}(\cdot)] \quad \text{for } i \in \underline{\nu}.$$

Objective: Construct a basis $(\psi_k(\cdot))_{k=1}^{n}$ of X, $n := \partial[\det D]$.

Procedure: For $i = 1 \sim \nu$ s.t. $n_i \neq 0$, do:

Step 1. Find n_i linearly independent solutions $\xi_i(\cdot) : [0, \infty) \to \mathbb{R}$ of the scalar differential equation

27 $$d_{ii}(p)\xi_i(t) = 0, \quad t \geq 0.$$

Step 2. If $i = 1$, skip step 2.
Else, for every linearly independent solution $\xi_i(\cdot)$ of step 1, for every $j = i-1, i-2, \cdots, 1$, solve for $\xi_j(\cdot)$ the scalar differential equation

28 $$d_{jj}(p)\xi_j(t) + d_{j(j+1)}\xi_{j+1}(t) + \cdots + d_{ji}(p)\xi_i(t) = 0$$

with initial conditions zero and driven by $\xi_i(\cdot), \xi_{i-1}(\cdot), \cdots, \xi_{j+1}(\cdot)$.
Step 3. From every independent solution $\xi_i(\cdot)$ of step 1 together with its corresponding set of solutions $\xi_j(\cdot)$ of step 2, (empty if $i = 1$), form a new basis vector of X given by

29 $$\psi_k(t) = \begin{bmatrix} \xi_1(t) \\ \vdots \\ \xi_{i-1}(t) \\ \xi_i(t) \\ \vdots \\ 0 \end{bmatrix} \in \mathbb{R}^\nu, \; t > 0-.$$

End of Algo.

30 Example. Consider the following example where n = 3.

$$D(p)\xi(t) = \begin{bmatrix} p+1 & p+1 \\ 0 & (p+2)^2 \end{bmatrix} \begin{bmatrix} \xi_1(t) \\ \xi_2(t) \end{bmatrix} = \theta_2 \quad t \geq 0.$$

$i = 1$: $n_1 = 1$: $\xi_1(t) = e^{-t}$,

$$\psi_1(t) = \begin{bmatrix} e^{-t} \\ 0 \end{bmatrix};$$

$i = 2$: $n_2 = 2$: (a) $\xi_2(t) = e^{-2t}$, $\xi_1(t) = e^{-t} - e^{-2t}$,

$$\psi_2(t) = \begin{bmatrix} e^{-t} - e^{-2t} \\ e^{-2t} \end{bmatrix};$$

(b) $\xi_2(t) = te^{-2t}$, $\xi_1(t) = -te^{-2t}$,

$$\psi_3(t) = \begin{bmatrix} -te^{-2t} \\ te^{-2t} \end{bmatrix}.$$

STOP. A basis for X is $(\psi_k(\cdot))_{k=1}^{3}$.

31 Remark. $n_i := \partial[d_{ii}(\cdot)] = 0$ means that no contribution is made to the solution space X: the solution of the scalar differential equation $d_{ii}(p)\xi_i(t) = 0$ $t \geq 0$ is $\xi_i(t) = 0$ for $t > 0-$; moreover, $\forall j = i-1, i-2, \cdots, 1$, the "driven" differential equation (28) is not driven: hence with initial conditions zero, $\forall j < i$, $\xi_j(t) = 0$ $\forall t > 0-$ $\cdots$.

32 Exercise. Consider the differential equation (1). Show that the reduction of $D(\cdot)$ to Hermite row form and the subsequent application of Algorithm 24 leads to the construction of a basis for the solution space X.

33 Remark. When applying the Laplace transform method in Algorithm 24 be sure to take into account the initial values of functions and their derivatives when needed.

34 Exercise. Compute a basis for the solution space X of the differential equation (1) where $D(p)$ is given by

$$D(p) = \begin{bmatrix} 1 & p & p^2 \\ 0 & p+1 & p \\ 0 & 0 & p^2 \end{bmatrix},$$

$$D(p) = \begin{bmatrix} 1 & p+1 & p^2 \\ 0 & 1 & p \\ 0 & 0 & 1 \end{bmatrix},$$

and

$$D(p) = \begin{bmatrix} (p+1)^2 & p \\ 1 & p+2 \end{bmatrix}.$$

An immediate consequences to Algorithm 24 is:

36 Algorithm [Construction of a state at t = 0 using initial data]
Data: we are given the differential equation (1) where $D(\cdot) \in \mathbb{R}[p]^{\nu\times\nu}$ is nonsingular and $n := \partial[\det D]$.
Step 1. By Algorithm 2.3.4.5.L reduce $D(\cdot)$ to the upper-triangular Hermite row form.
Step 2. For every $i \in \underline{\nu}$ compute n_i, the degree of entry (i, i) of the Hermite form.
Step 3. For every $i \in \underline{\nu}$ s.t. $n_i \neq 0$ compute the initial data

$$\xi_i(0),\ \xi_i^{(1)}(0),\ \cdots,\ \xi_i^{(n_i-1)}(0)$$

(denoting derivates of $\xi_i(\cdot)$ at 0), and stack them into an n-vector $x_\xi(0) \in \mathbb{R}^n$.

End of Algo.

37 Example. Consider the differential equation of Example 30: $n_1 = 1$, $n_2 = 2$; hence

$$x_\xi(0) = \begin{bmatrix} \xi_1(0) \\ \xi_2(0) \\ \xi_2^{(1)}(0) \end{bmatrix}.$$

(As an exercise show that $t \mapsto x_\xi(t)$ satisfies $\dot{x}_\xi = Ax_\xi$ for some $A \in \mathbb{R}^{3\times 3}$).

38 Theorem. The n-vector $x_\xi(0) \in \mathbb{R}^n$ constructed in Algorithm 36 is a state at $t = 0$ for the differential equation (1), or equiv. there exists a basis $(\bar{\psi}_k(\cdot))_{k=1}^n$ for the solution space X of (1) such that with the basis matrix $\bar{\Psi}(\cdot)$ (defined as in (13)),
(a) $\forall\ \xi(\cdot) \in X$, or equiv. $\forall$ solution of (1), there is a unique $x_\xi(0) \in \mathbb{R}^n$ s.t.

38a $$\xi(t) = \bar{\Psi}(t)x_\xi(0) \quad \text{for } t > 0-$$

(b) the map

38b $$x_\xi(0) \in \mathbb{R}^n \mapsto \xi(\cdot) = \bar{\Psi}(\cdot)x_\xi(0) \in X$$

is an $\mathbb{R}$-linear bijection (isomorphism) of $\mathbb{R}^n$ onto X. Moreover, let $(\psi_k(\cdot))_{k=1}^n$ be any basis of the solution space X of (1), with basis matrix $\psi(\cdot)$ defined by (13), and consider for differential equation (1) $x(0) \in \mathbb{R}^n$, viz., the state at $t = 0$ defined by equation (14).
U.t.c.
there exists a nonsingular matrix $P \in \mathbb{R}^{n\times n}$ s.t. the map

39 $$x_\xi(0) \in \mathbb{R}^n \mapsto x(0) = Px_\xi(0) \in \mathbb{R}^n$$

is a linear bijection. As a consequence every solution $\xi(\cdot)$ of (1) has the form

40 $$\xi(t) = \psi(t)Px_\xi(0) \quad \text{for } t > 0-.$$ ∎

41 Exercise. Prove Theorem 38. [Hints: (a) Use Algorithm 24 and the fact that the solution $\xi_i(\cdot)$ of any nontrivial scalar differential equation (27) is uniquely determined by $\xi_i(0), \xi_i^{(1)}(0), \cdots, \xi^{(n_i-1)}(0)$ with $n_i = \partial[d_{ii}(\cdot)]$; (b) use (19)-(20).]

42 Comment. The state $x_\xi(0) \in \mathbb{R}^n$ constructed by Algorithm 36 is specified by $\xi(0)$ and its derivatives evaluated at 0. For a precise formulation, see step 3 of Algorithm 36. Note in particular that if $n_i = 0$, $\xi_i(0)$ does not appear in $x_\xi(0)$.

At this point it is convenient to introduce the following definitions.

43 Definitions. Consider the differential equation (1). We call normalized state at $t = 0$ of (1) or simply the state at $t = 0$ of (1) the n-vector $x_\xi(0)$ constructed by Algorithm 36. We call (normalized) state trajectory of (1) the map $t \mapsto x_\xi(t)$. Finally, we call pseudo-state at $t = 0$ and pseudo-state trajectory of (1) $\xi(0)$ resp. the map $t \mapsto \xi(t)$.

44 Comments. (a) $x_\xi(0) = \theta_n$ is equivalent to $\xi(\cdot)$ and all its derivatives are zero.
(b) In general, $\xi(0) \neq x_\xi(0)$ and $\nu = \dim \xi(0) \neq n = \dim x_\xi(0) = \partial[\det D]$: we may have $\nu < n$ or $\nu = n$ or $\nu > n$.
(c) $\xi(t)$ is called pseudo-state because $\xi(0)$ and some of its derivates at 0 determine the state $x_\xi(0)$. In some cases the map $x_\xi(0) \mapsto \xi(0)$ is not surjective,(see (40) with $t = 0$).

45 Exercise. Consider differential equation (1) with $D(p) = pI - A$, where $A \in \mathbb{R}^{\nu\times\nu}$. (a) Show that $\nu = n = \partial[\det D]$ but in general $\xi(0) \neq x_\xi(0)$, (e.g., $A = \begin{bmatrix} 1 & 3 \\ 2 & 2 \end{bmatrix}$ leads to $x_\xi(0) = (\xi_2(0), \xi_2^{(1)}(0))^T$. (b) Knowing that $\xi(t) = \exp(At)\xi(0)$ for $t > 0-$, show that there exists a basis for X s.t. $x(0) = \xi(0)$, hence $\xi(0)$ is a state at $t = 0$.

We conclude this section with two important results. The first result characterizes trivial differential equations. The second result displays the invariance of the zero-state response of a PMD under e.o.'s.

50 Fact [Trivial dynamics]. Let $D(\cdot) \in \mathbb{R}[p]^{\nu\times\nu}$ be nonsingular and consider the differential equation

1 $$D(p)\xi(t) = \theta_\nu \quad t \geq 0$$

with solution space X.
U.t.c.

51 $$X = \{\theta\}$$

where θ stands for the solution $\xi(t) = \theta_\nu \ \forall t > 0-$,

$\Leftrightarrow$

$D(\cdot)$ is unimodular.

52 Exercise. Prove Fact 50.

54 Theorem [Invariance of the z-s response of a PMD under e.o.'s].
Consider the PMD $[D, N_\ell, N_r, K]$ described by

55
$$\begin{aligned} D(p)\xi(t) &= N_\ell(p)u(t) \\ y(t) &= N_r(p)\xi(t) + K(p)u(t) \end{aligned} \qquad t \geq 0$$

where (a) $D(\cdot)$, $N_\ell(\cdot)$, $N_r(\cdot)$, $K(\cdot) \in E(\mathbb{R}[p])$ and have dimensions $\nu \times \nu$, $\nu \times n_i$, $n_o \times \nu$, $n_o \times n_i$; moreover $D(\cdot)$ is nonsingular.

(b) $u(\cdot) : \mathbb{R}_+ \to \mathbb{R}^{n_i}$, $\xi(\cdot) : \mathbb{R}_+ \to \mathbb{R}^\nu$, $y(\cdot) : \mathbb{R}_+ \to \mathbb{R}^\nu$ are the input, pseudo-state, and output of the PMD.
(c) the z-s response $y(\cdot)$ is obtained by setting to zero the n-vector $x_\xi(0)$ (constructed in Algorithm 36), and driving the PMD by the input $u : \mathbb{R}_+ \to \mathbb{R}^{n_i}$, where by assumption $u(\cdot)$ and all its derivatives have zero value at $t = 0-$.

Let now

56
$$\begin{aligned} \bar{D}(p) &:= L(p)D(p)R(p) & \bar{N}_\ell(p) &:= L(p)N_\ell(p) \\ \bar{N}_r(p) &:= N_r(p)R(p) & \bar{K}(p) &:= K(p) \end{aligned}$$

where $L(\cdot)$ and $R(\cdot)$ are unimodular matrices representing e.o.'s over $\mathbb{R}[p]$. Consider then the transformed PMD $[\bar{D}, \bar{N}_\ell, \bar{N}_r, \bar{K}]$ described by

57
$$\left.\begin{aligned} \bar{D}(p)\bar{\xi}(t) &= \bar{N}_\ell(p)u(t), \\ y(t) &= \bar{N}_r(p)\bar{\xi}(t) + \bar{K}(p)u(t) \end{aligned}\right\} t \geq 0 .$$

U.t.c.

$\forall\, u(\cdot) : \mathbb{R}_+ \to \mathbb{R}^{n_i}$ piecewise sufficiently differentiable, the z-s responses of the PMD's (55) and (57) satisfy

$$y(t) = \bar{y}(t) \qquad \forall t > 0- .$$ ∎

1.L. Theorem [Hermite row form]. Let $M \in \mathbb{R}[s]^{m\times n}$. Then there exists a unimodular matrix $L \in \mathbb{R}[s]^{m\times m}$ (obtained by e.r.o.'s) such that

2 $\quad LM = H =$

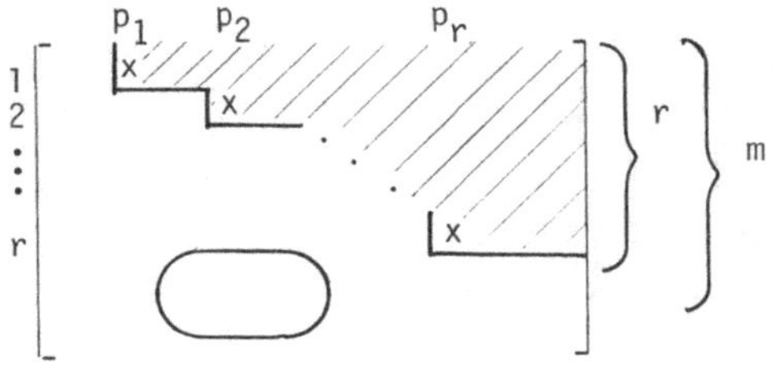

where H is called <u>Hermite row form of M</u> and has the following properties:

There exists an integer r, $0 \leq r \leq \min(m,n)$ with

3 $$\text{rk}H = \text{rk}M = r$$

s.t.

(a) $\forall i \in \underline{r}$, ρ_i has a leading nonzero <u>monic</u> polynomial h_{ip_i} called <u>leading entry</u> s.t.

$$1 \leq p_1 < p_2 < \cdots < p_r \leq n.$$

(b) $\forall i \in \underline{r}$,

if $h_{ip_i} = 1$, then $h_{jp_i} = 0 \quad \forall j < i$,

if $h_{ip_i} \neq 1$, then $\partial[h_{jp_i}] < \partial[h_{ip_i}] \quad \forall j < i \quad$ s.t. $h_{jp_i} \neq 0$.

(c) $\forall \gamma_j$ s.t. $j < p_1$, γ_j is zero;

$\forall \gamma_j$ s.t. $p_i \leq j < p_{i+1}$ with $i \in \underline{r-1}$, then the last m-i entries of γ_j are zero;

$\forall \gamma_j$ s.t. $j \geq p_r$, the last m-r entries of γ_j are zero. ∎

4 <u>Comment</u>. Conditions (a)-(c) above describe the <u>upper staircase</u> form of H with r stairs. Note the similarity of the Hermite row form with the <u>row echelon form</u> for matrices $M \in \mathbb{R}^{m\times n}$ [Nob.1]. The difference is that we cannot annihilate every entry above a leading entry: since we are working with polynomials, the Euclid algorithm can only decrease the degree of such

58 Exercise. Prove Fact 54. (Hint: Use the Laplace transform method where, at 0-, the values of $\xi(\cdot)$, $\bar{\xi}(\cdot)$, $u(\cdot)$, and their derivatives are zero by assumption.)

2.3.6. Greatest Common Divisor Extraction

We handle the case of a g.c.r.d.

1.R Algorithm [Greatest common right divisor extraction]

Data: $\bar{N}_r \in \mathbb{R}[s]^{n_o \times n_i}$, $\bar{D}_r \in \mathbb{R}[s]^{n_i \times n_i}$ with $\bar{D}_r$ nonsingular.

Step 1. Set

$$M := \begin{array}{c} \\ n_i \\ n_o \end{array}\!\!\begin{array}{c} n_i \\ \left[\begin{array}{c} \bar{D}_r \\ \hline \bar{N}_r \end{array}\right] \end{array} \in E(\mathbb{R}[s]) \tag{2}$$

and observe that M has full column rank over $\mathbb{R}[s]$.

Step 2. Use Algorithm 2.3.4.5.L to get M in upper triangular form by e.r.o.'s: see Corollary 2.3.4.9.L: step 6 of the algorithm may be skipped.

As a consequence we obtain unimodular matrices W and $W^{-1} \in E(\mathbb{R}[s])$ with

$$W := \left[\begin{array}{c|c} V_r & U_r \\ \hline -N_\ell & D_\ell \end{array}\right]\begin{array}{c} n_i \\ n_o \end{array} \qquad W^{-1} := \left[\begin{array}{c|c} D_r & -U_\ell \\ \hline N_r & V_\ell \end{array}\right]\begin{array}{c} n_i \\ n_o \end{array} \tag{3}$$

(columns: n_i, n_o)

such that

$$WM = \left[\begin{array}{c|c} V_r & U_r \\ \hline -N_\ell & D_\ell \end{array}\right]\left[\begin{array}{c} \bar{D}_r \\ \hline \bar{N}_r \end{array}\right]\begin{array}{c} n_i \\ n_o \end{array} = \left[\begin{array}{c} R \\ \hline 0 \end{array}\right]\begin{array}{c} n_i \\ n_o \end{array}, \tag{4}$$

where

R is upper-triangular and nonsingular. ∎

5 Remarks. (a) The unimodular matrices W and W^{-1} in (3) are obtained simultaneously by starting with a unit matrix $I_{n_o+n_i}$ for both W and W^{-1} and

performing upon it the e.r.o.'s of step 2 for getting W and their inverses as e.c.o.'s for getting W^{-1}: no inversion of W is needed.
The translation rules for e.r.o.'s and their inverses as e.c.o.'s are

1. $\rho_i \leftrightarrows \rho_j \quad : \quad \gamma_i \leftrightarrows \gamma_j$,
2. $\rho_i \leftarrow k\rho_i \quad : \quad \gamma_i \leftarrow k^{-1}\gamma_i$,
3. $\rho_i \leftarrow \rho_i + r\rho_j : \quad \gamma_j \leftarrow \gamma_j - r\gamma_i$.

(b) The reason for the special notation of the partitions of W and W^{-1} in (3) will become clear below.

6 Exercise. Apply Algorithm 1.R to

$$\bar{N}_r(s) = [(s+1)^2 \quad s-1] \;,\; \bar{D}_r(s) = \begin{bmatrix} s+1 & s-1 \\ 0 & (s-1)^2 \end{bmatrix},$$

and show that a result is

$$W = \left[\begin{array}{cc|c} -s & 1 & 1 \\ s+1 & -1 & -1 \\ \hline 1-s^2 & s & s-1 \end{array}\right], \quad W^{-1} = \left[\begin{array}{cc|c} 1 & 1 & 0 \\ 0 & s-1 & 1 \\ \hline s+1 & 1 & -1 \end{array}\right]$$

$$\begin{bmatrix} R \\ \hline 0 \end{bmatrix} = \left[\begin{array}{cc} s+1 & 0 \\ 0 & s-1 \\ \hline 0 & 0 \end{array}\right].$$

10.R Fact. Let $(\bar{N}_r, \bar{D}_r) \in \mathbb{R}[s]^{n_o \times n_i} \times \mathbb{R}[s]^{n_i \times n_i}$ with $\bar{D}_r$ nonsingular and apply Algorithm 1.R. Then the upper triangular matrix R in (4) is a g.c.r.d. of $(\bar{N}_r, \bar{D}_r)$.

Proof. (a) By (4)

$$W^{-1}\begin{bmatrix} R \\ \hline 0 \end{bmatrix} = \begin{bmatrix} \bar{D}_r \\ \hline \bar{N}_r \end{bmatrix}.$$

Hence by (3^2) $\bar{D}_r = D_r R$, $\bar{N}_r = N_r R$. Hence R is a c.r.d. of $(\bar{N}_r, \bar{D}_r)$.

(b) By (4) again, $U_r\bar{N}_r + V_r\bar{D}_r = R$. Hence any c.r.d. of $(\bar{N}_r, \bar{D}_r)$ is a r.d. of R, or equiv. R is a ℓ.m. of every c.r.d. of $(\bar{N}_r, \bar{D}_r)$. ∎

Recall now the definition of left-equivalence: see Sec. 2.3.2.

11.R Fact. Let $(\bar{N}_r, \bar{D}_r) \in \mathbb{R}[s]^{n_o \times n_i} \times \mathbb{R}[s]^{n_i \times n_i}$ with $\bar{D}_r$ nonsingular. Then all the g.c.r.d.'s of $(\bar{N}_r, \bar{D}_r)$ are left-equivalent, or equiv. if R' and R'' are g.c.r.d.'s of $(\bar{N}_r, \bar{D}_r)$; then

$$R' \overset{\ell}{\sim} R''.$$

12 Comment. All g.c.r.d.'s are related by e.r.o.'s or equiv. or a g.c.r.d. is unique up to a left unimodular factor.

Proof. Because left equivalence is an equivalence relation and by Fact 10.R matrix R of (4) is a g.c.r.d. of $(\bar{N}_r, \bar{D}_r)$, it is sufficient to show that

13 $$R' \underset{\sim}{\ell} R$$

for any R' a g.c.r.d. of $(\bar{N}_r, \bar{D}_r)$. Now since R and R' are g.c.r.d.'s of $(\bar{N}_r, \bar{D}_r)$, they are c.r.d.'s of $(\bar{N}_r, \bar{D}_r)$ and R must be a ℓ.m. of R' and vice versa, i.e., there exist square matrices L and $L' \in E(\mathbb{R}[s])$ s.t.

14 $$R' = L'R \qquad R = LR'.$$

Hence $R = LL'R$ with R nonsingular by (4). Hence $I = LL'$ such that L and L' are unimodular with $L' = L^{-1}$. Hence (13) holds. ∎

15.R Fact. Let $(N_r, D_r) \in \mathbb{R}[s]^{n_o \times n_i} \times \mathbb{R}[s]^{n_i \times n_i}$ with D_r nonsingular. Then

(N_r, D_r) is r.c.

⇔

all g.c.r.d.'s of (N_r, D_r) are unimodular.

Proof. Because of Fact 11.R if one g.c.r.d. is unimodular, then all g.c.r.d.'s are unimodular. ∎

We give now two characterizations of right-coprimeness.

16.R <u>Theorem</u> [Bezout identity]. Let $(N_r, D_r) \in \mathbb{R}[s]^{n_o \times n_i} \times \mathbb{R}[s]^{n_i \times n_i}$ with D_r nonsingular. Then

$$(N_r, D_r) \text{ is r.c.}$$

$\Leftrightarrow$

$$\exists\ U_r, V_r \in E(\mathbb{R}[s]) \text{ s.t.}$$

17
$$U_r N_r + V_r D_r = I_{n_i}.$$

18 <u>Comment</u>. Equation (17) is called the <u>Bezout identity</u> [Kai.1,p.379]. It generalizes to the multivariable case the well-known Bezout identity un + vd = 1 for coprime polynomials (n, d) [Sig.1]. Notice also that condition (17) shows that $[V_r\ U_r]$ is a left inverse of the matrix $\begin{bmatrix} D_r \\ \hdashline N_r \end{bmatrix}$ where all operations are within the ring $\mathbb{R}[s]$.

19 <u>Proof of Theorem 16.R</u>. (a) $\Rightarrow$: Use Algorithm 1.R with $\bar{N}_r \leftarrow N_r$ and $\bar{D}_r \leftarrow D_r$. Then one obtains by (4) $U_r\bar{N}_r + V_r\bar{D}_r = R$ with R a g.c.r.d. of $(\bar{N}_r, \bar{D}_r)$ by Fact 10.R. Now $(\bar{N}_r, \bar{D}_r)$ is a right coprime pair. Hence, by Fact 15.R, R must be unimodular. Therefore, by premultiplication by R^{-1} one obtains $\bar{U}_r\bar{V}_r + \bar{V}_r\bar{D}_r = I_{n_i}$, where $\bar{U}_r = R^{-1}U_r$ and $\bar{V}_r = R^{-1}V_r \in E(\mathbb{R}[s])$: we have a Bezout identity.
(b) $\Leftarrow$: Because of (17) any g.c.r.d. of (N_r, D_r) is unimodular. ∎

20.R <u>Theorem</u> [Rank test]. Let $(N_r, D_r) \in \mathbb{R}[s]^{n_o \times n_i} \times \mathbb{R}[s]^{n_i \times n_i}$ with D_r nonsingular.
U.t.c.

$$(N_r, D_r) \text{ is r.c.}$$

$\Leftrightarrow$

21
$$\operatorname{rk} \overset{n_i}{\begin{bmatrix} D_r(s) \\ \hdashline N_r(s) \end{bmatrix}} \begin{matrix} n_i \\ \\ n_o \end{matrix} = n_i \qquad \forall s \in \mathbb{C}.$$

22 <u>Comment</u>. Condition (21) is known as the <u>rank test</u>.

(a) It generalizes to the multivariable case the well-known condition that a polynomial pair (n, d) is coprime iff they have no common zeros, i.e., iff $rk[d(s)\ n(s)]^T = 1$, $\forall s \in \mathbb{C}$, i.e., iff they have no noninvertible common factors. For our matrix case this reads: (N_r, D_r) is r.c. iff they have no nonunimodular right common factors.

(b) From Comment (18) it follows also: the full column rank matrix $\begin{bmatrix} D_r \\ \hline N_r \end{bmatrix}$ is left invertible over $\mathbb{R}[s]$ iff the rank test (21) holds.

23 Exercise. Use the rank test (21) to show that $(\bar{N}_r, \bar{D}_r)$ as given in Exercise 6 is not r.c. Find a nonunimodular right common factor.

24 Proof of Theorem 20.R. Apply Algorithm 1.R with $\bar{N}_r \leftarrow N_r$ and $\bar{D}_r \leftarrow D_r$. According to (4) and Fact 10.R one obtains

$$W\begin{bmatrix} \bar{D}_r \\ \hline \bar{N}_r \end{bmatrix} = \begin{bmatrix} R \\ \hline 0 \end{bmatrix},$$

where (a) W is unimodular and represents e.r.o.'s; (b) R is a g.c.r.d. of $(\bar{N}_r, \bar{D}_r)$. Observe now that e.r.o.'s result in rank invariance at every $s \in \mathbb{C}$ and that by Fact 15.R (N_r, D_r) is r.c. iff R is unimodular. Hence

$$rk\begin{bmatrix} D_r(s) \\ N_r(s) \end{bmatrix} = n_i, \ \forall s \in \mathbb{C} \ \Leftrightarrow \ R \text{ is unimodular } \Leftrightarrow (N_r, D_r) \text{ is r.c.}$$ ∎

27 Convention. Algorithm 1.R extracts a g.c.r.d. $R \in \mathbb{R}[s]^{n_i \times n_i}$ from $(\bar{N}_r, \bar{D}_r) \in \mathbb{R}[s]^{n_o \times n_i} \times \mathbb{R}[s]^{n_i \times n_i}$ with $\bar{D}_r$ nonsingular. Consider now $(\bar{D}_\ell, \bar{N}_\ell) \in \mathbb{R}[s]^{n_o \times n_o} \times \mathbb{R}[s]^{n_o \times n_i}$ with $\bar{D}$ nonsingular. Then similarly by e.c.o.'s on

28
$$M = \overset{\quad n_o \quad\ n_i}{[\bar{D}_\ell \,\vdots\, \bar{N}_\ell]}\, n_o$$

one can get M in lower triangular form by Algorithm 2.3.4.5.R and Corollary 2.3.4.9.R: see Convention 2.3.4.12. As a consequence there exists a unimodular matrix $W \in E(\mathbb{R}[s])$ s.t.

29 $$MW =: \left[\bar{D}_\ell \;\vdots\; \bar{N}_\ell\right] \overset{\quad n_o \quad\; n_i}{\begin{bmatrix} V_\ell & -N_r \\ U_\ell & D_r \end{bmatrix}} = \begin{bmatrix} L & 0 \end{bmatrix} n_o ,$$

where

L is <u>lower-triangular</u> and <u>nonsingular</u>.

End of Procedure

The procedure above is called Algorithm 1.L. [Extraction of a g.c.ℓ.d.]. Its consequences are similar to those of Algorithm 1.R, viz., Fact 10.L (L in (29) is a g.c.ℓ.d. of $(\bar{D}_\ell, \bar{N}_\ell)$). Fact 11.L (all g.c.ℓ.d.'s of $(\bar{D}_\ell, \bar{N}_\ell)$ are right-equivalent), Fact 15.L ($(\bar{D}_\ell, \bar{N}_\ell)$ is ℓ.c. iff every g.c.ℓ.d. is unimodular), Theorem 16.L (Bezout identity for left coprimeness), Theorem 20.L (rank test for left coprimeness). Notice that we have used R (resp. L) to indicate a result on right extraction, right coprimeness···(resp. left extraction, left coprimeness): we adopt the following convention:

Throughout the text a suffix R (resp. L) will mean that we are referring to a procedure or result which has a <u>dual analog</u> indicated by a suffix L (resp. R) which should be recognizable from the context.

30 <u>Exercise</u>. Draw a comparative table with in the first column Facts 10.R, 11.R, 15.R, Theorems 16.R, 20.R and in the second column Facts 10.L, 11.L, 15.L, Theorems 16.L, 20.L.

2.4. <u>Matrix Fraction Descriptions of Rational Transfer Function Matrices</u>

In this section we consider rational transfer functions as polynomial matrix fractions.

2.4.1. <u>Coprime Fractions</u>

1.R <u>Definitions</u>. Let $H \in \mathbb{R}(s)^{n_o \times n_i}$. We say that $(N_r, D_r) \in \mathbb{R}[s]^{n_o \times n_i} \times \mathbb{R}[s]^{n_i \times n_i}$ is a <u>right coprime fraction</u> (r.c.f.) of H iff

(a) $\det D_r \not\equiv 0$,
(b) $H = N_r D_r^{-1}$,
(c) (N_r, D_r) is r.c.

If (c) is not required we say that (N_r, D_r) is a right fraction (r.f.) of H. ∎

1.L Definitions. Let $H \in \mathbb{R}(s)^{n_o \times n_i}$. We say that $(D_\ell, N_\ell) \in \mathbb{R}[s]^{n_o \times n_o} \times \mathbb{R}[s]^{n_o \times n_i}$ is a left coprime fraction (ℓ.c.f.) of H iff

(a) $\det D_\ell \not\equiv 0$,

(b) $H = D_\ell^{-1} N_\ell$,

(c) (D_ℓ, N_ℓ) is ℓ.c.

If (c) is not required, we say that (D_ℓ, N_ℓ) is a left fraction (ℓ.f.) of H. ∎

In the following we consider in detail right fractions; we leave it to the reader to investigate left fractions.

2.R Algorithm [Search of a r.c.f.]

Data: $H \in \mathbb{R}(s)^{n_o \times n_i}$.

Step 1. $\forall j \in \underline{n}_i$ compute

3 d_j := a least common denominator (ℓ.c.d.) of all entries of γ_j, where

$d_j := 1$ if γ_j is zero.

Step 2. $\forall i \in \underline{n}_o$ and $\forall j \in \underline{n}_i$ write every entry h_{ij} of H as

4 $$h_{ij} = \bar{n}_{ij}/d_j .$$

Set

5 $$\bar{N}_r := [\bar{n}_{ij}],\ \bar{D}_r := \operatorname{diag}[d_j]_{j=1}^{n_i} .$$

Comment: $(\bar{N}_r, \bar{D}_r)$ is a r.f. of H.

Step 3. Using Algorithm 2.3.6.1.R, extract a g.c.r.d. $R \in \mathbb{R}[s]^{n_i \times n_i}$ of $(\bar{N}_r, \bar{D}_r)$. As a result:

6 $$\bar{N}_r = N_r R \quad \bar{D}_r = D_r R.$$

Comment: (N_r, D_r) is a r.c.f. of H.

End of Algo

7 Remarks. (a) Steps 1 and 2 may be replaced by any method which delivers a r.f. $(\bar{N}_r, \bar{D}_r)$ of H.
(b) The application of extraction Algorithm 2.3.6.1.R in step 3 delivers matrices N_r, D_r of a r.c.f. as submatrices of the partition of unimodular matrix W^{-1} in (2.3.6.3). See the proof of Fact 8.R.

8.R Fact. Every $H \in \mathbb{R}(s)^{n_o \times n_i}$ has a r.c.f. (N_r, D_r).

Proof. Use Algorithm 2.R and observe that in step 3 (6) holds since (2.3.6.3) and (2.3.6.4) result in

$$W^{-1}\begin{bmatrix} R \\ \hline 0 \end{bmatrix} = \begin{bmatrix} \bar{D}_r \\ \hline \bar{N}_r \end{bmatrix}.$$

Moreover, since $WW^{-1} = I$, a Bezout identity is obtained from the partitions of W and W^{-1} in (2.3.6.3), viz.,

$$U_r N_r + V_r D_r = I_{n_i}.$$

Hence, according to Theorem 2.3.6.16.R, (N_r, D_r) is r.c.. It follows that (N_r, D_r) is a r.c.f. of H. ∎

Algorithm 2.R does more than producing a r.c.f. of H•••.

9.R Fact. Let $H \in \mathbb{R}(s)^{n_o \times n_i}$ and apply Algorithm 2.R. The result of Algorithm 2.R is that, at the end of step 3, one obtains from the submatrices of W and W^{-1} in (2.3.6.3):
(a)

(N_r, D_r) is a r.c.f. of H

with, since $WW^{-1} = I$, an associated Bezout identity:

$$U_r N_r + V_r D_r = I_{n_i}.$$

(b)

(D_ℓ, N_ℓ) is a ℓ.c.f. of H

with, since $WW^{-1} = I$, an associated Bezout identity:

$$N_\ell U_\ell + D_\ell V_\ell = I_{n_o}.$$

10 Comment. Algorithm 2.R produces simultaneously a r.c.f. and a ℓ.c.f. of H. Fact 9.R explains also the notation used in the partitions of unimodular matrices W and W^{-1} in (2.3.6.3) at the end of extraction Algorithm 2.3.6.1.R: the eight polynomial matrices N_r, D_r, U_r, V_r; N_ℓ, D_ℓ, U_ℓ, V_ℓ are data for 2 coprime fractions of the same rational matrix.

11 Proof of Fact 9.R. (a) follows by the the proof of Fact 8.R.
(b) From (2.3.6.3) with $WW^{-1} = I$, one has

12 $$N_\ell U_\ell + D_\ell V_\ell = I_{n_o}$$

13 $$N_\ell D_r = D_\ell N_r$$

where by (a) $H = N_r D_r^{-1}$.

Hence we are done if we show that

$$\det D_\ell \not\equiv 0.$$

This is done by contradiction. Assume that $\exists\, \eta \in \mathbb{R}[s]^{n_o}$, η nonzero, s.t. $\eta^T D_\ell \equiv \theta$, where θ stands for a row vector which is zero. Then by (13) and $\det D_r \not\equiv 0$, $\eta^T N_\ell \equiv \theta$. Hence by (12),

$$\theta \equiv \eta^T (N_\ell U_\ell + D_\ell V_\ell) = \eta^T \not\equiv \theta : \rightarrow\leftarrow.$$ ■

14 Exercise. Use the data and results of extraction Exercise 2.3.6.6.
(a) Show that $(\bar{N}_r, \bar{D}_r)$ is a r.f. of

$$H(s) = [s + 1 \,\vdots\, -s(s - 1)^{-1}].$$

(b) Using step 3 of Algorithm 2.R, obtain a r.c.f. and a ℓ.c.f. of H with their associated Bezout identities from (2.3.6.3). ■

We conclude this section with a result on uniqueness and the description of the Generalized Bezout Identity of any rational matrix.

17.R Theorem [Uniqueness of r.c.f.'s]. Let $H \in \mathbb{R}(s)^{n_o \times n_i}$. Then any two r.c.f.'s (N_{r1}, D_{r1}) and (N_{r2}, D_{r2}) of H are equal modulo a common unimodular right factor, or equiv.

$\exists$ a common unimodular matrix $R \in \mathbb{R}[s]^{n_i \times n_i}$, (representing e.c.o.'s on $\begin{bmatrix} D_{r1} \\ N_{r1} \end{bmatrix}$), s.t.

18 $$N_{r2} = N_{r1}R \qquad D_{r2} = D_{r1}R.$$

Proof. For i = 1, 2 (N_{ri}, D_{ri}) is r.c. Therefore, by the Bezout Identity Theorem 2.3.6.16.R for i = 1, 2 $\exists\ U_{ri}, V_{ri} \in E(\mathbb{R}[s])$ s.t.

19 $$U_{ri}N_{ri} + V_{ri}D_{ri} = I \quad \text{for } i = 1, 2.$$

Moreover, for i = 1, 2 $H = N_{ri}D_{ri}^{-1}$, whence

20 $$N_{r2} = N_{ri}D_{r1}^{-1}D_{r2}.$$

Set now

21 $$R := D_{r1}^{-1}D_{r2}.$$

From (20) and (21), (18) is established; it remains to be shown that R in (21) is unimodular. Now from (19) with i = 1, $U_{r1}N_{r2} + V_{r1}D_{r2} = R$; hence R is a polynomial matrix. From (18) $N_{r1} = N_{r2}R^{-1}$, $D_{r1} = D_{r2}R^{-1}$. Hence by (19) with i = 2, $U_{r2}N_{r1} + V_{r2}D_{r1} = R^{-1}$; hence R^{-1} is also a polynomial matrix and therefore unimodular. ∎

22 Exercise. State and prove Theorem 17.L [Uniqueness of ℓ.c.f.'s].

25.R Theorem [Generalized Bezout identity generated by a r.c.f.]. Let $H \in \mathbb{R}(s)^{n_o \times n_i}$ have a r.c.f. (N_r, D_r). Then there exist six matrices $\in E(\mathbb{R}[s])$, viz.,

$U_r, V_r, N_\ell, D_\ell, U_\ell, V_\ell$

s.t.

27
$$W\,W^{-1} := \begin{array}{c} \\ n_i \\ n_o \end{array}\!\!\begin{array}{c} \begin{array}{cc} n_i & n_i \end{array} \\ \left[\begin{array}{c|c} V_r & U_r \\ \hline -N_\ell & D_\ell \end{array}\right] \end{array} \begin{array}{c} \begin{array}{cc} n_i & n_o \end{array} \\ \left[\begin{array}{c|c} D_r & -U_\ell \\ \hline N_r & V_\ell \end{array}\right] \end{array}\!\!\begin{array}{c} \\ n_i \\ n_o \end{array} = \left[\begin{array}{c|c} I_{n_i} & O \\ \hline O & I_{n_o} \end{array}\right].$$

Moreover,

28 (D_ℓ, N_ℓ) is a ℓ.c.f. of H.

29 Comment. Equation (27) is called a generalized Bezout identity for H, e.g., [Kai.1,p.382]. It is a key tool: see below. The shaded area in (27) shows known data.

30 Proof of Theorem 25.R. Apply extraction Algorithm 2.3.6.1.R with $\bar{N}_r \leftarrow N_r$, $\bar{D}_r \leftarrow D$. Note that because of (2.3.6.4)

31
$$WM := \left[\begin{array}{c|c} V_r & U_r \\ \hline -N_\ell & D_\ell \end{array}\right]\left[\begin{array}{c} D_r \\ \hline N_r \end{array}\right] = \left[\begin{array}{c} R \\ \hline 0 \end{array}\right],$$

where, by Fact 2.3.6.10.R, R is a g.c.r.d. of (N_r, D_r). Hence by Fact 2.3.6.15.R, R is unimodular, because (N_r, D_r) is r.c. As a consequence we may set $R = I_{n_i}$ (it suffices to multiply in (31) (block ρ_1) on the left by R^{-1}: note that $R^{-1}V_rD_r + R^{-1}U_rN_r = I_{n_i}$). Now, by a proof similar to the proof of Fact 2.4.1.9.R, (D_ℓ, N_ℓ) in (31) is a ℓ.c.f. of H, whence by Theorem 2.3.6.16.R $\exists\ \tilde{U}_\ell, \tilde{V}_\ell \in E(\mathbb{R}[s])$ s.t.

$$N_\ell\tilde{U}_\ell + D_\ell\tilde{V}_\ell = I_{n_o}.$$

As a result one obtains from the above:

$$\left[\begin{array}{c|c} V_r & U_r \\ \hline -N_\ell & D_\ell \end{array}\right]\left[\begin{array}{c|c} D_r & -\tilde{U}_\ell \\ \hline N_r & \tilde{V}_\ell \end{array}\right] = \left[\begin{array}{c|c} I_{n_i} & Q \\ \hline O & I_{n_o} \end{array}\right],$$

where $Q := -V_r\tilde{U}_\ell + U_r\tilde{V}_\ell$. Finally, using block-column operations, viz.,

$$(\text{block } \gamma_2) \leftarrow (\text{block } \gamma_2) - (\text{block } \gamma_1)\, Q,$$

we obtain (27), defining

$$-U_\ell := \tilde{U}_\ell - D_r Q \qquad V_\ell := \tilde{V}_\ell - N_r Q.$$

We are done since (28) has already been established. ∎

32 Exercise. State Theorem 25.L [Generalized Bezout identity generated by a ℓ.c.f.].

2.4.2. Smith-McMillan Form; Relation to Coprime Fractions

The Smith-McMillan form of a rational transfer function matrix is a conceptual tool. It clarifies the relations between all coprime fractions of a rational matrix. In Sec. 2.4.4 it will be used to define poles and zeros of such a matrix.

1 Theorem [Smith-McMillan form]. Let $H \in \mathbb{R}(s)^{n_o \times n_i}$ have normal rank r. Then there exist unimodular matrices $L \in \mathbb{R}[s]^{n_o \times n_o}$ and $R \in \mathbb{R}[s]^{n_i \times n_i}$ (obtained by e.o.'s over the polynomial ring $\mathbb{R}[s]$), s.t.

2 $$H = LMR,$$

where

3 $$M := \begin{bmatrix} \text{diag}[(\varepsilon_i/\psi_i)]_{i=1}^{r} & 0 \\ 0 & 0 \end{bmatrix} \begin{matrix} r \\ n_o - r \end{matrix} \in \mathbb{R}(s)^{n_o \times n_i},$$
$$\qquad\qquad\qquad r \qquad\qquad n_i - r$$

with

4 (ε_i, ψ_i) a pair of monic coprime polynomials for $i \in \underline{r}$,

5 $\psi_{i+1} | \psi_i$ for $i \in \underline{r-1}$,

6 $\varepsilon_i | \varepsilon_{i+1}$ for $i \in \underline{r-1}$,

7 $\psi_1 = d :=$ the monic ℓ.c.d. of all entries of H.

M is called the Smith-McMillan form of H. ■

8 Comment. The only entries of the Smith-McMillan form, (SMM-form), that are nonzero are the diagonal entries of the r-dimensional top left block with $r = \text{rk}H$; the latter entries are rational functions.

9 Proof of Theorem 1. With d := the monic ℓ.c.d. of all entries of H, we can write every entry of H as

10 $$h_{ij} = n_{ij}/d,$$

where n_{ij} and d are polynomials. Therefore, by defining

11 $$N := [n_{ij}] \in \mathbb{R}[s]^{n_o \times n_i},$$

one obtains

12 $$dH = N,$$

where N is a polynomial matrix. Hence, according to Theorem 2.3.4.18, N has a unique Smith form S and $\exists$ unimodular matrices L and R s.t.

13 $$dH = N = LSR,$$

where

14 $$S = \begin{bmatrix} \text{diag}[\lambda_i]_{i=1}^{r} & 0 \\ 0 & 0 \end{bmatrix} \begin{matrix} r \\ n_o - r \end{matrix}, \qquad \text{column widths } r,\ n_i - r$$

15 $$r = \text{rk}S = \text{rk}H,$$

16 $$\lambda_i | \lambda_{i+1} \quad \text{for } i \in \underline{r-1}.$$

Finally, after dividing (13) by d:

17 $$H = LMR,$$

where

18 $$M = S\,d^{-1} = \left[\begin{array}{c|c} \operatorname{diag}[(\varepsilon_i/\psi_i)]_{i=1}^{r} & 0 \\ \hline 0 & 0 \end{array}\right]\begin{matrix} r \\ n_o - r \end{matrix},$$
$$\qquad\qquad\qquad\quad r \qquad\qquad n_i - r$$

because after canceling common factors,

19 $$(\lambda_i/d) =: (\varepsilon_i/\psi_i) \in \mathbb{R}(s) \quad \text{for} \quad i \in \underline{r},$$

with

20 (ε_i,ψ_i) a pair of monic coprime polynomials for $i \in \underline{r}$.

Hence we have proved (2)-(4).

Division properties (5) and (6) are consequences of (16), (19), and (20). Indeed, $\lambda_{i+1}/\lambda_i = (\varepsilon_{i+1}\ \psi_i)/(\varepsilon_i\ \psi_{i+1})$ is a polynomial where $(\varepsilon_i,\ \psi_i)$ and $(\varepsilon_{i+1},\ \psi_{i+1})$ are coprime polynomials; therefore, $\varepsilon_i | \varepsilon_{i+1}$ and $\psi_{i+1} | \psi_i$.

Property (7) is established by contradiction. Assume therefore that $\psi_1 \neq d$:= the monic least common denominator of all entries of H. Then, because $(\varepsilon_1/\psi_1) = (\lambda_1/d)$ with $(\varepsilon_1,\ \psi_1)$ coprime, the polynomials λ_1 and d must have a common factor (one gets $(\varepsilon_1,\ \psi_1)$ coprime by canceling common factors in (λ_1/d)). Now, by the Smith form Theorem 2.3.4.18, $\lambda_1 = \Delta_1$ = the monic g.c.d. of <u>all</u> entries of polynomial matrix N (see (10)-(16)). Hence d and <u>all</u> entries of N have a common factor. This contradicts the fact that d is the <u>least</u> common denominator of every entry of H. ∎

Recall that by Theorems 2.4.1.17.R and 2.4.1.17.L all r.c.f.'s (ℓ.c.f.'s) of a given rational matrix $H \in \mathbb{R}(s)^{n_o \times n_i}$ are equal modulo a common right (resp. left) unimodular factor. A convenient way to relate arbitrary coprime fractions (right and/or left) is by using the SMM-form. Therefore, consider $H \in \mathbb{R}(s)^{n_o \times n_i}$ and its SMM-form as described in Theorem 1 by (2)-(7).

Define now the following polynomial matrices based on (2)-(7):

25 $$\boldsymbol{\mathcal{E}} := \begin{bmatrix} \operatorname{diag}[\varepsilon_i]_{i=1}^{r} & \bigcirc \\ \bigcirc & \bigcirc \end{bmatrix} \begin{matrix} r \\ n_o - r \end{matrix}, \qquad \text{columns: } r,\ n_i - r$$

26 $$\boldsymbol{\Psi}_r := \begin{bmatrix} \operatorname{diag}[\psi_i]_{i=1}^{r} & \bigcirc \\ \bigcirc & I \end{bmatrix} \begin{matrix} r \\ n_i - r \end{matrix}, \qquad \text{columns: } r,\ n_i - r$$

27 $$\boldsymbol{\Psi}_\ell := \begin{bmatrix} \operatorname{diag}[\psi_i]_{i=1}^{r} & \bigcirc \\ \bigcirc & I \end{bmatrix} \begin{matrix} r \\ n_o - r \end{matrix}, \qquad \text{columns: } r,\ n_o - r$$

28 $$N_r := L\boldsymbol{\mathcal{E}}, \quad D_r := R^{-1}\boldsymbol{\Psi}_r,$$

29 $$D_\ell := \boldsymbol{\Psi}_\ell L^{-1}, \quad N_\ell := \boldsymbol{\mathcal{E}}R.$$

Note that (a) the <u>dimensions</u> of $\boldsymbol{\Psi}_r$ and $\boldsymbol{\Psi}_\ell$ have been <u>adapted</u> to match n_i := the # of inputs and n_o := the # of outputs, and (b) that $\boldsymbol{\Psi}_r$ and $\boldsymbol{\Psi}_\ell$ are in <u>reverse Smith form</u> $(\psi_i | \psi_{i-1})$.

Notice now that by Theorem 1 and the definitions above

30 $$H = N_r D_r^{-1} = D_\ell^{-1} N_\ell,$$

where at every $s \in \mathbb{C}$

31 $$\operatorname{rk}\begin{bmatrix} D_r \\ N_r \end{bmatrix} = \operatorname{rk}\underbrace{\begin{bmatrix} R^{-1} & \bigcirc \\ \bigcirc & L \end{bmatrix}}_{\text{unimodular}}\begin{bmatrix} \boldsymbol{\Psi}_r \\ \boldsymbol{\mathcal{E}} \end{bmatrix} = \operatorname{rk}\begin{bmatrix} \boldsymbol{\Psi}_r \\ \boldsymbol{\mathcal{E}} \end{bmatrix} = n_i,$$

and similarly at every $s \in \mathbb{C}$

32 $$\operatorname{rk}[D_\ell \mid N_\ell] = \operatorname{rk}[\Psi_\ell \ \ \mathcal{E}] = n_o .$$

From (30) we have that (N_r, D_r) $((D_\ell, N_\ell))$ is a r.f. (resp. ℓ.f.) of $H \in \mathbb{R}(s)^{n_o \times n_i}$. By rank test Theorems 2.3.6.20.R and 2.3.6.20.L it follows that the pairs (N_r, D_r) and (D_ℓ, N_ℓ) are coprime. Hence

33 Fact. The SMM-form generates through matrices (25)-(29) a r.c.f. (N_r, D_r) and a ℓ.c.f. (D_ℓ, N_ℓ) for any $H \in \mathbb{R}(s)^{n_o \times n_i}$. ∎

We shall now relate arbitrary coprime fractions of $H \in \mathbb{R}(s)^{n_o \times n_i}$ using the notion of equivalence (see Sec. 2.3.2), and uniqueness Theorems 2.4.1.17.R and 2.4.1.17.L.

36 Theorem [Numerators of coprime fractions]. Let $H \in \mathbb{R}(s)^{n_o \times n_i}$. Then all the numerators of any coprime fraction of H (right and/or left) are equivalent and have the same Smith form $\mathcal{E}$ given by (25).

37 Comment. Numerators can be obtained from each other by e.o.'s: they have the same normal rank and the same local rank at every $s \in \mathbb{C}$.

38 Proof of Theorem 36. By the uniqueness Theorems 2.4.1.17.R and 2.4.1.17.L we have only to investigate the relation between numerators N_r and N_ℓ of Fact 33. Now, by (28)-(29) with L and R unimodular, $N_r \sim \mathcal{E} \sim N_\ell$. Therefore, equivalence follows with as unique Smith form, matrix $\mathcal{E}$ given by (25): notice that $\varepsilon_i \mid \varepsilon_{i+1}$. ∎

41 Theorem [Denominators of coprime fractions]. Let $H \in \mathbb{R}(s)^{n_o \times n_i}$. Then all the denominators of any coprime fraction of H (right and/or left) have the same nonunity invariant polynomials. Hence, in particular, their determinants are equivalent (equiv. equal modulo a nonzero constant). Moreover, if $n_o = n_i$, (equiv. H is square), then these denominators are equivalent.

42 Comment. In the nonsquare case there is no equivalence because right and left denominators have different dimensions.

43 Exercise. Prove Theorem 41. (Hint: Use the uniqueness Theorems 2.4.1.17.R and 2.1.4.17.L and observe that (28)-(29) $D_\ell \sim \Psi_\ell$, $D_r \sim \Psi_r$, while in (26)-(27) Ψ_ℓ and Ψ_r have the same nonunity invariant polynomials.)

44 Exercise. Let $H \in \mathbb{R}(s)^{n_o \times n_i}$, where $n_o > n_i$. Let (N_r, D_r) be a r.c.f. and (D_ℓ, N_ℓ) be a ℓ.c.f. of H. Show that

$$D_\ell \sim \left[\begin{array}{c|c} D_r & 0 \\ \hline 0 & I_{n_o - n_i} \end{array}\right].$$

2.4.3. Proper Transfer Function Matrices

Most physical models have low-pass frequency characteristics, hence finite gain as the frequency is increased···. For linear time-invariant systems described by a transfer function this means that the latter has to be bounded at infinity.

1 Definition. Let $H \in \mathbb{R}(s)^{n_o \times n_i}$. Then we say that H is proper (strictly proper) iff $\lim_{s\to\infty} H(s) = H(\infty) \in \mathbb{C}^{n_o \times n_i}$ ($\lim_{s\to\infty} H(s) = 0$, resp.). This is denoted by $H \in \mathbb{R}_p(s)^{n_o \times n_i}$ ($H \in \mathbb{R}_{p,o}(s)^{n_o \times n_i}$, resp.), where $\mathbb{R}_p(s)$ ($\mathbb{R}_{p,o}(s)$) is the ring of proper (strictly proper) rational functions with coefficients in $\mathbb{R}$.

∎

2 Exercise. Show that $\mathbb{R}_p(s)$ and $\mathbb{R}_{p,o}(s)$ are subrings of the field $\mathbb{R}(s)$.

3 Definition. Let $m \in \mathbb{R}[s]^n$ be a polynomial vector. We define the degree of m, denoted by $\partial[m]$, to be the highest degree of all entries of the vector. If the vector m is row i or column j of the polynomial matrix $M \in \mathbb{R}[s]^{n_o \times n_i}$, then their degrees are called ith row-degree or jth column-degree and denoted by $\partial_{ri}[M]$, $\partial_{cj}[M]$ resp. Note that for $m \in \mathbb{R}[s]^n$, $\partial[m] = -\infty$ iff m is the zero vector.

A rational function is proper if and only if the degree of the denominator is at least the degree of the numerator. In the matrix case this is only necessary column-wise or row-wise. (However, see 25.R below.)

4.R Fact. Let $H \in \mathbb{R}_p(s)^{n_o \times n_i}$ ($\in \mathbb{R}_{p,o}(s)^{n_o \times n_i}$) have a r.f. (N_r, D_r); then

5 $\quad \forall j \in \underline{n}_i \quad \partial_{cj}[N_r] \le \partial_{cj}[D_r]$

6 $\quad (\forall j \in \underline{n}_i \quad \partial_{cj}[N_r] < \partial_{cj}[D_r]$, resp.).

Proof. We prove the proper case and drop subscripts r. Note that N = HD with $N = [n_{ij}]$, $D = [D_{ij}]$ reads

$$n_{ij} = \sum_{k=1}^{n_i} h_{ik} d_{kj}.$$

Let now $k_j := \partial_{cj}[D]$; then taking limits we have

$$\lim_{s\to\infty} n_{ij}(s)\, s^{-k_j} = \sum_{k=1}^{n_i} (\lim_{s\to\infty} h_{ik}(s))\,(\lim_{s\to\infty} d_{kj}(s)\, s^{-k_j}),$$

where all limits on the RHS exist and are finite. Hence the limit on the LHS must exist and be finite. As a consequence, $\forall i \in \underline{n}_o, \partial[n_{ij}] \le k_j := \partial_{cj}[D]$. Hence taking the maximum over i in the LHS, (5) follows. ∎

7 Exercise. State and prove Fact 4.L. (Let $H \in \mathbb{R}_p(s)^{n_o \times n_i}$ have a ℓ.f. (D_ℓ, N_ℓ) then $\forall i \in \underline{n}_o \ \partial_{ri}[N_\ell] \le \cdots$.)

8 Remark. the converse of Fact 4.R is not true in general. For example, with

$$N_r(s) = [s \,\vdots\, 1], \quad D_r(s) = \begin{bmatrix} 1 & 1 \\ s & s^2 \end{bmatrix}$$

condition (5) is satisfied, but $H(s) = \begin{bmatrix} \dfrac{s^3-s}{s^2-s} & \dfrac{1-s}{s^2-s} \end{bmatrix} \notin \mathbb{R}_p(s)^{1\times 2}$.

In the analysis below a condition is added to make the converse true$\cdots$.

11 Definition. Let $D \in \mathbb{R}[s]^{n\times n}$ be a nonsingular polynomial matrix. We say that D is column-reduced (c.r.) (resp. row-reduced (r.r.)) iff

$\partial[\det D] = \sum_{j=1}^{n} \partial_{cj}[D]$ ($\partial[\det D] = \sum_{i=1}^{n} \partial_{ri}[D]$, resp.).

12 Exercise. Show that one always has

13 $$\partial[\det D] \le \sum_{j=1}^{n} \partial_{cj}[D].$$

(Hint: Set $k_j := \partial_{cj}[D]$ and consider $\lim_{s\to\infty} (\det D(s) (s^{-\sum_{j=1}^{n} k_j}))$.) ∎

14 Remark. Exercise 12 suggests that column-reducedness can be achieved by using e.c.o.'s to successively reduce the individual column-degrees until column-reducedness is achieved. For example, the e.c.o. $\gamma_2 \leftarrow \gamma_2 - s\gamma_1$ will transform

$$D(s) = \begin{bmatrix} 1 & 1 \\ s & s^2 \end{bmatrix} \text{ into } \begin{bmatrix} 1 & 1-s \\ s & 0 \end{bmatrix}, \text{ which is c.r.}$$

See also, e.g., [Kai.1, Example 6.3-2]. As a consequence, if $H \in \mathbb{R}(s)^{n_o \times n_i}$ has a r.f. $(\bar{N}_r, \bar{D}_r)$, then by e.c.o.'s on matrix $\begin{bmatrix} \bar{D}_r \\ \bar{N}_r \end{bmatrix}$, H will have a r.f. (N_r, D_r) where D_r is c.r.; moreover, if $(\bar{N}_r, \bar{D}_r)$ is a r.c.f., then (N_r, D_r) remains a r.c.f. Similar statements can be made for row-reducedness. Hence without loss of generality we may assume that the denominator of a (coprime) fraction is column- or row-reduced.

15 Highest Column-degree Coefficient Matrix. For any nonsingular polynomial matrix $D \in \mathbb{R}[s]^{n\times n}$, let

16 $$k_j := \partial_{cj}[D] \quad \text{for } j \in \underline{n};$$

then

17 $$\lim_{s\to\infty} D(s)\, (\mathrm{diag}[s^{-k_j}]_{j=1}^{n}) =: D_h \in \mathbb{R}^{n\times n},$$

where $D_h \in \mathbb{R}^{n\times n}$ is the highest column-degree coefficient matrix of D (representing the coefficients of s^{k_j} of each entry d_{ij} of D). Moreover, since the determinant is continuous in its arguments

18 $$\lim_{s\to\infty} \left(\det D(s) \left(s^{-\sum_{j=1}^{n} k_j}\right)\right) = \det D_h \in \mathbb{R}\,. \qquad ■$$

We have then, by Definition 11 and (18),

21.R <u>Fact</u>. Let $D \in \mathbb{R}[s]^{n\times n}$ be a nonsingular polynomial matrix with column-degrees k_j and highest column-degree coefficient matrix $D_h \in \mathbb{R}^{n\times n}$.

U.t.c.

D is column-reduced (c.r.)

⇔

D_h is nonsingular

⇔

$$\lim_{s\to\infty} D(s)\,(\mathrm{diag}[s^{-k_j}]_{j=1}^{n}) = D_h, \text{ with } D_h \text{ nonsingular.} \qquad ■$$

22 <u>Exercise</u>. State Fact 21.L (row-reducedness, row-degrees, highest row-degree coefficient matrix).

Fact 4.R can now be refined to an equivalence.

25.R <u>Theorem</u>. Let $H \in \mathbb{R}(s)^{n_o\times n_i}$ admit a r.f. (N_r, D_r) with D_r c.r..

U.t.c.

$$H \in \mathbb{R}_p(s)^{n_o\times n_i}, \quad (H \in \mathbb{R}_{p,o}(s)^{n_o\times n_i})$$

⇔

26 $$\forall j \in \underline{n}_i \qquad \partial_{cj}[N_r] \le \partial_{cj}[D_r]$$

27 $$(\forall j \in \underline{n}_i \qquad \partial_{cj}[N_r] < \partial_{cj}[D_r], \text{ resp.}).$$

28 <u>Comment</u>. Modulo a column-reduced denominator the characterization of a proper right fraction is a straightforward extension of the scalar case.

29 <u>Proof of Theorem 25.R</u>. ⇒ : follows by Fact 4.R.

⇐ : Drop subscripts r. Then, using the column-degrees k_j of D as given by (16), one obtains

$$H(s) = (N(s)\,\mathrm{diag}[s^{-k_j}])\,(D(s)\,\mathrm{diag}[s^{-k_j}])^{-1},$$

where

(a) $\lim_{s\to\infty} D(s)\,\mathrm{diag}[s^{-k_j}] = D_h$, with D_h nonsingular since D is c.r. and by Fact 21.R;

(b) $\lim_{s\to\infty} N(s)\,\mathrm{diag}[s^{-k_j}] =: N_h \in \mathbb{R}^{n_o\times n_i}$ (= 0, resp.) because of (26), ((27) resp.).

Hence $\lim_{s\to\infty} H(s) = N_h D_h^{-1}$ (= 0, resp.). ∎

Two other properties are important; we state the first.

32.R <u>Theorem</u> [Uniqueness of column-degrees], e.g., [Kai.1, Lemma 6.3.14]. Let $D \in \mathbb{R}[s]^{n\times n}$ and $\bar{D} \in \mathbb{R}[s]^{n\times n}$ be two <u>column-reduced</u> nonsingular polynomial matrices with column-degrees arranged in the same order (increasing or decreasing). U.t.c., if D and $\bar{D}$ are <u>right equivalent</u>, then their column-degrees are <u>identical</u>.

33.R <u>Comment</u>. Theorem 32.R shows that if $H \in \mathbb{R}(s)^{n_o\times n_i}$ has a r.c.f. (N_r, D_r) with D_r column-reduced and column-degrees in increasing or decreasing order, then these <u>column-degrees are a property of H</u>: all r.c.f.'s with the properties mentioned above are right equivalent and therefore display the same column-degrees.

34 <u>Exercise</u>. State Theorem 32.L. [Uniqueness of row-degrees]. What about Comment 33.L?

37.R <u>Theorem</u> [Division Theorem], e.g., [Kai.1, Theorem 6.3.15]. Let $D_r \in \mathbb{R}[s]^{n_i\times n_i}$ be a nonsingular polynomial matrix. Then for any polynomial matrix $N_r \in \mathbb{R}[s]^{n_o\times n_i}$ there exist <u>unique</u> polynomial matrices $Q_r \in \mathbb{R}[s]^{n_o\times n_i}$ and $R_r \in \mathbb{R}[s]^{n_o\times n_i}$ such that

38 $$N_r = Q_r D_r + R_r \quad \text{with} \quad R_r D_r^{-1} \in \mathbb{R}_{p,o}(s)^{n_o\times n_i}.$$

Moreover, if D_r is also c.r., then the uniqueness of Q_r and R_r will also be ensured iff

39 $$\partial_{cj}[R_r] < \partial_{cj}[D_r] \quad \forall j \in \underline{n}_i.$$

Proof. Observe that $H := N_r D_r^{-1} \in \mathbb{R}(s)^{n_o \times n_i}$; hence

$$\forall i,j \quad h_{ij} := n_{ij}/d_{ij} \in \mathbb{R}(s),$$

where (n_{ij}, d_{ij}) is a coprime polynomial pair. Therefore, by the Euclid algorithm $\forall i,j$ $\exists!$ polynomials q_{ij}, r_{ij} s.t.

$$n_{ij} = d_{ij} q_{ij} + r_{ij} \quad \text{with} \quad r_{ij}/d_{ij} \in \mathbb{R}_{p,o}(s).$$

Hence

$$\forall i,j \quad h_{ij} = q_{ij} + r_{ij}/d_{ij},$$

s.t. with

$$Q_r := [q_{ij}] \in \mathbb{R}[s]^{n_o \times n_i} \text{ and } H_{sp} := [r_{ij}/d_{ij}] \in \mathbb{R}_{p,o}(s)^{n_o \times n_i}$$

$$N_r D_r^{-1} = H = Q_r + H_{sp}.$$

Hence with

$$R_r := N_r - Q_r D_r \in \mathbb{R}[s]^{n_o \times n_i}.$$

(38) follows since $R_r D_r^{-1} = H_{sp} \in \mathbb{R}_{sp}(s)^{n_o \times n_i}$. By Theorem 25.R the latter holds. The uniqueness of the pair (Q_r, R_r) holds because, if $(\bar{Q}_r, \bar{R}_r)$ is another pair such that (38) holds, then, from $N_r = Q_r D_r + R_r = \bar{Q}_r D_r + \bar{R}_r$,

$$Q_r - \bar{Q}_r = (\bar{R}_r - R_r) D_r^{-1},$$

where the LHS $\in \mathcal{E}(\mathbb{R}[s])$ and the RHS $\in \mathcal{E}(\mathbb{R}_{p,o}(s))$. Now $\mathbb{R}[s] \cap \mathbb{R}_{p,o}(s) = \{0\}$. Hence $Q_r = \bar{Q}_r$ and $R_r = \bar{R}_r$. ∎

40 Exercise. State Theorem 37.L [Division on the left]. As a final comment it should be stressed that division can be performed by an algorithm.

41 Exercise. (a) Study the algorithmic division on the left of a polynomial vector by a nonsingular polynomial matrix in Appendix C.

(b) Let

$$N_\ell(s) := \left[\begin{array}{c|c} s^4 & s^3 \\ \hline s^2 & s^3 \end{array}\right], \quad D_\ell(s) := \left[\begin{array}{c|c} s^3 + 4s^2 + 5s + 2 & s + 2 \\ \hline s + 1 & s^2 + 4s + 4 \end{array}\right].$$

Show that the quotient Q_ℓ and remainder R_ℓ of the division on the left of N_ℓ by D_ℓ is given by

$$Q_\ell(s) := \left[\begin{array}{c|c} s - 4 & 1 \\ \hline 0 & s - 4 \end{array}\right], \quad R_\ell(s) = \left[\begin{array}{c|c} 11s^2 + 18s + 8 & -5s^2 - 3s + 6 \\ \hline 3s + 4 & 11s + 15 \end{array}\right].$$

(Hint for (b): The division on the left of N_ℓ by D_ℓ is obtained by successively dividing the columns of N_ℓ by D_ℓ.)

2.4.4. Poles and Zeros

1. Definitions. Let $H \in \mathbb{R}(s)^{n_o \times n_i}$ and consider its SMM-form $M \in \mathbb{R}(s)^{n_o \times n_i}$ (Theorem 2.3.2.1), given by

$$M = \left[\begin{array}{c|c} \operatorname{diag}[(\varepsilon_i/\psi_i)]_{i=1}^{r} & \bigcirc \\ \hline \bigcirc & \bigcirc \end{array}\right] \begin{array}{l} r \\ n_o - r \end{array}, \qquad \text{columns: } r, \; n_i - r$$

where, for $i \in \underline{r}$, ε_i and ψ_i are monic coprime polynomials having properties (2.4.2.5)-(2.4.2.7) and where r is the normal rank of H.

We call pole of H any root of any denominator polynomial ψ_i, $i \in \underline{r}$, of the SMM-form of H. The set of poles of H is denoted by $P[H]$.

We call zero of H any root of any numerator polynomial ε_i, $i \in \underline{r}$, of the SMM-form of H. The set of zeros of H is denoted by $Z[H]$. ∎

2 Comment. Note that a pole of H may also be a zero of H, e.g., let $M = \operatorname{diag}\{\frac{1}{s+1}, 1, (s+1)\}$. So for the multivariable case, poles and zeros may coincide: this is due to the fact that they have value in $\mathbb{C}$ and "place in the matrix."

3 Fact. Let $H \in \mathbb{R}(s)^{n_o \times n_i}$ with $H = [h_{ij}]_{i \in \underline{n}_o, j \in \underline{n}_i}$. Then

(a) $p \in P[H] \Leftrightarrow \exists\, h_{ij}$ s.t. $p \in P[h_{ij}]$;

(b) $p \notin P[H] \Leftrightarrow$ the map $s \mapsto H(s)$ in bounded in $N(p)$, where $N(p)$ denotes a sufficiently small neighborhood of $p \in \mathbb{C}$.

4 Comments. (a) means that at least one entry of H has a pole at p.
(b) Since the map $s \mapsto H(s)$ is meromorphic in s (equiv. has $\forall p \in \mathbb{C}$ a Laurent expansion in some $N(p)$ which stops at most at a negative power of $(s-p)$), the map $s \mapsto H(s)$ is bounded in some $N(p)$ iff the map $s \mapsto H(s)$ is analytic at p.

5 Proof of Fact 3. (a) Recall Definition 1 and the fact that $\psi_i | \psi_{i-1}$; consequently, $p \in P[H]$ if and only if $\psi_1(p) = d(p) = 0$, where d is the ℓ.c.d. of the entries of H. Hence $p \in P[H]$ iff $\exists\, h_{ij}$ s.t. $p \in P[h_{ij}]$.
(b) By the negation of (a):

$$p \notin P[H] \Leftrightarrow \forall i,j,\ p \notin P[h_{ij}]$$

$\Leftrightarrow \forall i,j$, map $s \mapsto h_{ij}(s)$ is bounded in some $N_{ij}(p)$

$\Leftrightarrow \forall i,j$, map $s \mapsto H(s)$ is bounded in some $N(p)$

(take the intersection of the $N_{ij}(p)$'s above). ∎

We shall now characterize poles and zeros in terms of any coprime fraction of a rational transfer function matrix. Recall therefore the developments of Sec. 2.4.2 on the SMM-form and its relation to coprime fractions. To avoid unnecessary redundancy in the transfer function system description, we shall also assume that the normal rank of $H \in \mathbb{R}(s)^{n_o \times n_i}$ is $r = \min(n_o, n_i)$, (otherwise, there are "trivial inputs or outputs").

6 Theorem [Poles and zeros of coprime fractions]. Let $H \in \mathbb{R}(s)^{n_o \times n_i}$ have normal rank $r = \min(n_o, n_i)$. Let (N_r, D_r), resp. (D_ℓ, N_ℓ), be any r.c.f. or ℓ.c.f. of H.
U.t.c.
(a)

$$p \in P[H] \Leftrightarrow \det D_r(p) = 0 \Leftrightarrow \det D_\ell(p) = 0$$

(b)

$$z \in Z[H]$$

$\Leftrightarrow$

$$\mathrm{rk}[N_r(z)] < r = \min(n_o, n_i)$$

$\Leftrightarrow$

$$\mathrm{rk}[N_\ell(z)] < r = \min(n_o, n_i).$$

7 Proof of Theorem 6. (a) According to denominator Theorem 2.4.2.41 and Fact 2.4.2.33 using (2.4.2.25)-(2.4.2.29) we have

$$\det D_\ell \sim \det D_r \sim \det \Psi_r \sim \det \Psi_\ell.$$

Now according to Definitions 1 we have

$$p \in P[H] \Leftrightarrow \det \Psi_r(p) = 0 \Leftrightarrow \det \Psi_\ell(p).$$

Hence assertion (a) is true.

(b) According to Numerator Theorem 2.4.2.36 and Fact 2.4.2.33, using (2.4.2.25)-(2.4.2.29) we have $N_\ell \sim N_r \sim \mathcal{E}$. Now according to Definitions 1 we have $z \in Z[H] \Leftrightarrow \mathrm{rk}[\mathcal{E}(z)] < r$. Hence assertion (b) is true. ∎

An immediate consequence of Theorem 6 is that zeros are poles of the inverse transfer function matrix, and vice versa.

8 Fact. Let $H \in \mathbb{R}(s)^{n \times n}$ be nonsingular. Then

$$z \in Z[H] \Leftrightarrow p \in P[H^{-1}]; \; p \in P[H] \Leftrightarrow p \in Z[H^{-1}].$$

9 Exercise. Prove Fact 8. (Hint: If (N_r, D_r) is a r.c.f. of H, then (D_r, N_r) is a r.c.f. of $\cdots$.)

Recall now that a state space system description (SSD) is a quadruple [A, B, C, D] of real matrices describing $\dot{x} = Ax + Bu$, $y = Cx + Du$ with A, B, C, D $\in \mathbb{R}^{n \times n}$, $\mathbb{R}^{n \times n_i}$, $\mathbb{R}^{n_o \times n}$, $\mathbb{R}^{n_o \times n_i}$, resp.. Any SSD [A, B, C, D] defines a transfer function $H(s) = C(sI - A)^{-1} B + D \in \mathbb{R}_p(s)^{n_o \times n_i}$, where without loss of generality $\mathrm{rk}[C \mid D] = n_o$ and $\mathrm{rk}\begin{bmatrix} B \\ \hline D \end{bmatrix} = n_i$ (otherwise, there are "trivial outputs or inputs"). A minimal realization of $H(s) \in \mathbb{R}_p(s)^{n_o \times n_i}$ is a SSD [A, B, C, D] s.t. A has minimal dimension. We have then

12 Theorem [Poles and zeros of minimal realizations]. Let $H \in \mathbb{R}_p(s)^{n_o \times n_i}$

and let $[A, B, C, D]$ be any minimal realization of order $n = \partial[\det(sI - A)]$, with $\operatorname{rk}\begin{bmatrix} B \\ \hline D \end{bmatrix} = n_i$ and $\operatorname{rk}[C \mid D] = n_o$.

U.t.c.

(a)

13 $$p \in \mathcal{P}[H] \;\Leftrightarrow\; \det(pI - A) = 0;$$

(b) if

14 $$P(s) := \begin{array}{c} \\ n \\ n_o \end{array}\!\!\begin{array}{c} \begin{array}{cc} n & n_i \end{array} \\ \left[\begin{array}{c|c} sI - A & B \\ \hline -C & D \end{array}\right] \end{array}$$

denotes the system matrix [Ros.1], then

15 $$z \in Z[H] \;\Leftrightarrow\; \operatorname{rk}[P(z)] < n + \min(n_o, n_i).$$ ∎

16 Proof of Theorem 12. Since (A, B) and (C, A) are controllable resp. observable, $(sI - A)^{-1}B$ is a ℓ.c.f. and $C(sI - A)^{-1}$ is a r.c.f.. Therefore, by Theorems 2.4.1.25L and 2.4.1.25R we have two generalized Bezout identities, viz.,

17 $$\left[\begin{array}{c|c} \bar{V}_r & \bar{U}_r \\ \hline -B & sI - A \end{array}\right]\left[\begin{array}{c|c} \bar{D}_r & -\bar{U}_\ell \\ \hline \bar{N}_r & \bar{V}_\ell \end{array}\right] = \left[\begin{array}{c|c} I & 0 \\ \hline 0 & I \end{array}\right],$$

18 $$\left[\begin{array}{c|c} \tilde{V}_r & \tilde{U}_r \\ \hline -\tilde{N}_\ell & \tilde{D}_\ell \end{array}\right]\left[\begin{array}{c|c} sI - A & -\tilde{U}_\ell \\ \hline C & \tilde{V}_\ell \end{array}\right] = \left[\begin{array}{c|c} I & 0 \\ \hline 0 & I \end{array}\right],$$

where

19 $$(\bar{N}_r, \bar{D}_r) \text{ is a r.c.f. of } (sI - A)^{-1}B$$

and where by Theorem 2.4.2.41

20 $$\det(sI - A) \sim \det \bar{D}_r.$$

Claim 1: $(C\bar{N}_r, \bar{D}_r)$ is a r.c.f. of $C(sI - A)^{-1}B$.

By (19) it is sufficient to show that $(C\bar{N}_r, \bar{D}_r)$ is r.c. Now from the generalized Bezout identities (17)-(18) one has

21 $$\tilde{V}_r(sI - A) + \tilde{U}_rC = I,$$

22 $$\bar{V}_r\bar{D}_r + \bar{U}_r\bar{N}_r = I.$$

On multiplying (21) on the right by $(sI - A)^{-1}B$ and using (19), we get successively

$$\tilde{V}_rB + \tilde{U}_rC(sI - A)^{-1}B = (sI - A)^{-1}B,$$
$$\tilde{V}_rB + \tilde{U}_rC\bar{N}_r\bar{D}_r^{-1} = \bar{N}_r\bar{D}_r^{-1},$$
$$\tilde{V}_rB\bar{D}_r + \tilde{U}_rC\bar{N}_r = \bar{N}_r.$$

On substituting this expression for $\bar{N}_r$ in (22), we get finally

$$(\bar{V}_r + \bar{U}_r\tilde{V}_rB)\bar{D}_r + (\bar{U}_r\tilde{U}_r)C\bar{N}_r = I.$$

This is a Bezout identity for $(C\bar{N}_r, \bar{D}_r)$, which is therefore r.c. ∎

Claim 2: $(C\bar{N}_r + D\bar{D}_r, \bar{D}_r)$ is a r.c.f. of H.

The proof of this claim is left as an exercise.

Claim 3: The system matrix $P(\cdot)$ defined in (14) is equivalent to

$$\left[\begin{array}{c|c} I_n & 0 \\ \hline 0 & C\bar{N}_r + D\bar{D}_r \end{array}\right].$$

Indeed, using (17) (where the matrices on the LHS are unimodular), and starting by the product of $P(\cdot)$ and a unimodular matrix we obtain successively

$$\left[\begin{array}{c|c} sI - A & B \\ \hline -C & D \end{array}\right]\left[\begin{array}{c|c} \bar{V}_\ell & \bar{N}_r \\ \hline \bar{U}_\ell & -\bar{D}_r \end{array}\right] = \left[\begin{array}{c|c} I_n & 0 \\ \hline -C\bar{V}_\ell + D\bar{U}_\ell & -C\bar{N}_r - D\bar{D}_r \end{array}\right]$$

$$\sim \left[\begin{array}{c|c} I_n & 0 \\ \hline 0 & C\bar{N}_r + D\bar{D}_r \end{array}\right].$$

Thus claim 3 follows. ∎

<u>Conclusion</u>. Using Claim 2, Theorem 6, (20), and Claim 3:

(a) $p \in \mathcal{P}[H] \Leftrightarrow \det \bar{D}_r(p) = 0 \Leftrightarrow \det(pI - A) = 0$

(b) $z \in Z[H] \Leftrightarrow rk[C\bar{N}_r + D\bar{D}_r](z) < \min(n_o, n_i)$

$$\Leftrightarrow rk[P(z)] < n + \min(n_o, n_i).$$

End of Proof

The following properties of zeros are consequences of Theorem 12.

25 <u>Fact</u>. Let [A, B, C, 0] be any minimal realization of $H \in \mathbb{R}_{p,o}(s)^{n_o \times n_i}$ with $rk[B] = n_i$ and $rk[C] = n_o$. Then using characterization (15), the zeros of H are <u>invariant</u> under

(a) <u>algebraic equivalence</u> $(A \leftarrow TAT^{-1}, B \leftarrow TB, C \leftarrow CT^{-1}, \det T \neq 0)$,

(b) <u>constant output feedback</u> $(A \leftarrow A + BKC, B \leftarrow B, C \leftarrow C)$. ∎

26 <u>Exercise</u>. Prove Fact 25 (Hint: Note that minimality is conserved because algebraic equivalence and constant output feedback upset neither controllability nor observability, e.g., [Che.1]; use also (15).)

To motivate the next theorem consider the following:

27 <u>Exercise</u>. Let

$$h(s) := ((s - 1)(s - 2))/((s + 1)(s + 2)(s + 3)) \in \mathbb{R}_{p,o}(s)$$

and consider the controllable, resp. observable, canonical realizations [A, b, c, o] of h, e.g., [Che.1, Theorems 7-1 and 7-2].

(a) Find a state feedback vector $k \in \mathbb{R}^{1 \times n}$, representing gains from each state to the input of the controllable realization, such that the numerator of the resulting transfer function becomes a <u>nonzero constant</u> (equiv. all the zeros of h(s) have been canceled).

(b) Find a feedback vector $k \in \mathbb{R}^{n\times 1}$, representing gains from the output to each integrator input of the observable realization, such that the numerator of the resulting transfer function becomes a nonzero constant. ∎

Exercise 27 shows that for the scalar case pole-zero cancellation is associated with (a) suitable constant state to input feedback ($A \leftarrow A + BK$, $B \leftarrow B$, $C \leftarrow C$) and (b) suitable constant output to integrator-input feedback ($A \leftarrow A + KC$, $B \leftarrow B$, $C \leftarrow C$). Note that in the literature (a) constant state to input feedback is known as state feedback, e.g., [Che.1, Sec. 7-3], and (b) constant output to integrator-input feedback is related to state reconstruction by observers, e.g., [Che.1, Sec. 7-4]. This leads to the following characterization of zeros in the multivariable case.

28 Theorem [Pole-zero cancellation under constant state to input feedback and constant output to integrator-input feedback]. Let $H \in \mathbb{R}_{p,o}(s)^{n_o\times n_i}$ with a minimal realization $[A, B, C, 0]$ and $rk(B) = n_i$, $rk(C) = n_o$

(a) Let $n_i \le n_o$; then

29
$$z \in Z[H] \Leftrightarrow \begin{cases} \exists \text{ a nonzero vector } x_0 \in \mathbb{C}^n, \ \exists \text{ a matrix } K \in \mathbb{C}^{n_i\times n} \\ \text{s.t. } (A + BK)x_0 = zx_0,\ Cx_o = \theta \end{cases}$$

or equiv., for any zero z, there is a constant state to input feedback matrix $K \in \mathbb{C}^{n_i\times n}$ such that for the resulting system $[A + BK, B, C, 0]$ the zero z is a system eigenvalue which is unobservable.

(b) Let $n_o \le n_i$; then

30
$$z \in Z[H] \Leftrightarrow \begin{cases} \exists \text{ a nonzero vector } x_0 \in \mathbb{C}^n, \ \exists \text{ a matrix } K \in \mathbb{C}^{n\times n_o} \\ \text{s.t. } (A + KC)^T x_0 = zx_0,\ B^T x_0 = \theta, \end{cases}$$

or equiv., for any zero z, there is a constant output to integrator-input matrix $K \in \mathbb{C}^{n\times n_o}$ such that for the resulting system $[A + KC, B, C, 0]$ the zero z is a system eigenvalue which is uncontrollable.

31 Comment. Note, e.g., [Che.1, Secs. 7-3 and 7-4] that under state to input feedback (output to integrator-input feedback), controllability (observability

resp.) is maintained, so minimality can only be lost through loss of observability, (controllability resp.). Note also that (29) means that $\mathrm{rk}\begin{bmatrix} zI - A - BK \\ \hline C \end{bmatrix} < n$, whence the pair $(C, sI - A - BK)$ is not r.c.; hence it has a common nonunimodular right factor R s.t. $\det R(z) = 0$. As a result a pole-zero cancellation occurs in the transfer function $C(sI - A - BK)^{-1}B$. This transfer function has the same zeros as $H(s) = C(sI - A)^{-1}B$ iff minimality is conserved. A similar pole-zero cancellation occurs in $C(sI - A - KC)^{-1}B$ under (30).

32 Proof of Theorem 28. (a) Observe that $\left[\begin{array}{c|c} zI - A & B \\ \hline -C & 0 \end{array}\right]\begin{bmatrix} x_0 \\ \hline -Kx_o \end{bmatrix} = \theta_{n+n_o}$, whence by (15) $z \in Z[H]$.

$\Rightarrow$: By assumption $z \in Z[H]$, whence by (15) $\exists$ a nonzero vector $(x_0^T, u_0^T)^T$ s.t.

33 $$\left[\begin{array}{c|c} zI - A & B \\ \hline -C & 0 \end{array}\right]\begin{bmatrix} x_0 \\ \hline u_0 \end{bmatrix} = \theta_{n+n_o}$$

Now $\mathrm{rk}(B) = n_i$ and (C, A) is observable. Hence both x_0 and u_0 are nonzero; furthermore, B has a left inverse B^{ℓ} s.t. $B^{\ell}B = I_{n_i}$. Hence from the first equation of (33), $B^{\ell}(zI - A)x_0 + u_0 = \theta$. So with $K := B^{\ell}(zI - A)$ we have from (33) $(A + BK)x_0 = zx_0$ and $Cx_o = \theta$.

(b) is left as an exercise. ∎

We conclude this section with a short consideration of poles and zeros at infinity.

35 Definition. [Kai.1, p.449]. Let $H \in \mathbb{R}(s)^{n_o \times n_i}$ and consider the map $s \in \mathbb{C} \mapsto \lambda = s^{-1} \in \mathbb{C}$; hence $s = \infty$ iff $\lambda = 0$. The poles and zeros of H at infinity are by definition the poles and zeros of the map $\lambda \in \mathbb{C} \mapsto H(\lambda^{-1})$ at $\lambda = 0$. ∎

Note that for observing the pole-zero structure at infinity we make a change of variables $s = \lambda^{-1}$ and investigate the SMM-form of the rational matrix $H(\lambda^{-1})$ at $\lambda = 0$.

36.R Exercise. Prove that if $D \in \mathbb{R}[s]^{n \times n}$ is a nonsingular column-reduced polynomial matrix, then D has no zero at ∞. (Hint: Consider $D(\lambda^{-1})$ and Theorem 6....)

37 Exercise. Prove: let $H \in \mathbb{R}(s)^{n_o \times n_i}$; then H is proper if and only if H has no pole at ∞. (Hint: Consider $H(\lambda^{-1})$ and a reasoning similar to that of Fact 3.)

38 Exercise. Prove: let $D \in \mathbb{R}[s]^{n\times n}$ be a nonsingular polynomial matrix; then D has no zeros at ∞ if and only if D^{-1} is proper. (Hint: A pole at ∞ of D^{-1} is a zero at ∞ of D⋯.)

39 Fact [Pug.1, Thm 5]. Let $D \in \mathbb{R}[s]^{n\times n}$ be a nonsingular polynomial matrix; then D has no zeros at ∞ if and only if

40 $$\delta[\det D] = \delta_M[D],$$

where

$\delta_M[D]$ denotes the McMillan degree of the polynomial matrix D [Ros.1, pp.134-135, p. 117], and equals the maximal degree of any minor of any order of D.

41 Exercise. Show that

$$D(s) = \begin{bmatrix} 1 & s^3 & 0 \\ 0 & s & s \\ 0 & 0 & s^2 \end{bmatrix}$$

violates criterion (40). Hence D has a zero at ∞.

42 Remark. Criterion (40) is not very practical since it involves calculating the degrees of all minors of any order of the polynomial matrix D. Now in Remark 2.4.3.14 we mentioned that any nonsingular $D \in \mathbb{R}[s]^{n\times n}$ can be converted in a column-reduced matrix D_1 by e.c.o.'s ⋯. By Exercise 36 D_1 has no zeros at ∞, or equiv. D_1^{-1} is proper. This suggests a more practical criterion for having D^{-1} proper⋯.

43.R Fact. Let $D \in \mathbb{R}[s]^{n\times n}$ be a nonsingular polynomial matrix and let $R \in \mathbb{R}[s]^{n\times n}$ be any unimodular matrix (obtained by e.c.o.'s) such that

44 $$DR = D_1,$$

where D_1 is c.r..

U.t.c.

45 D has no zeros at ∞ (equiv. D^{-1} is proper)

if and only if

46 $$\forall j \in \underline{n} \quad \partial_{cj}[R] \leq \partial_{cj}[D_1].$$

47 Comment. "If R is cheap, then criterion (46) is cheap."

48 Exercise. Show that the matrix D of Exercise 41 does not satisfy criterion (46).

49 Exercise. Prove Fact 43.R (Hint: Use $RD_1^{-1} = D^{-1}$ and Theorem 2.4.3.25.R. $\cdots$.)

50 Exercise. Get Fact 43.L (conversion to a row-reduced matrix, bounds on the row-degrees of the transforming matrix).

2.4.5. Dynamical Interpretation of Poles and Zeros

In this section we consider time-domain interpretations of transfer function-related definitions. The Laplace transform will be indicated by a superscript ^ or the symbol $\mathcal{L}[\cdots]$. The symbol $\mathcal{L}^{-1}[\cdots]$ is used for indicating the inverse Laplace transform.

1 Theorem [Poles]. Consider the transfer function matrix $\hat{H} \in \mathbb{R}(s)^{n_o \times n_i}$. U.t.c.
If $p \in P[\hat{H}]$, then $\exists$ an input $u(\cdot)$ which is a distribution with support at t=0 given by

2 $$u(t) = \sum_{k=0}^{\ell} u_k \, \delta^{(k)}(t), \text{ with } u_k \in \mathbb{C}^{n_i}, \ \forall k = o \sim \ell$$

(whence

3 $$\hat{u}(s) = \sum_{k=0}^{\ell} u_k \, s^k \in \mathbb{C}[s]^{n_i}),$$

s.t. the output $\hat{y}(s) = \hat{H}(s)\hat{u}(s)$ satisfies

4 $$y(t) = \gamma \exp(pt) + \eta(t) \quad \forall t \geq 0$$

with γ a <u>nonzero</u> vector of $\mathbb{C}^{n_o}$, $\hat{\eta}(s) \in \mathbb{C}[s]^{n_o}$. ∎

5 <u>Comment</u>. The input $u(\cdot)$ "kicks" the system into the correct initial condition at $t = 0+$ such that, for $t > 0$, $y(t) = \gamma \exp(pt)$ (indeed, $\eta(\cdot)$ is a distribution with support at $t = 0$). "The input $u(\cdot)$, defined by (2), excites <u>only</u> the exponential mode p."

6 <u>Proof of Theorem 1</u>. Let (N_r, D_r) be a r.c.f. of $\hat{H}$. Since p is a pole of a <u>coprime</u> fraction, $\det D_r(p) = 0$ and $\exists$ a nonzero vector $\xi \in \mathbb{C}^{n_i}$ s.t. $D_r(p)\xi = \theta$; moreover, $[N_r(p)^T \mid D_r(p)^T]^T$ has full column rank and so $\gamma := N_r(p)\xi \neq \theta$. Set now $\hat{u}(s) := D_r(s)\xi/(s - p)$ and observe that $\hat{u} \in \mathbb{C}[s]^{n_i}$. Moreover, $\hat{y}(s) = N_r(s)\xi/(s - p) = \gamma/(s - p) + (N_r(s) - N_r(p))\xi/(s - p)$, where $\hat{\eta}(s) := (N_r(s) - N_r(p))\xi/(s - p) \in \mathbb{C}[s]^{n_o}$. ∎

From now on <u>we assume that the transfer function matrix $\hat{H} \in \mathbb{R}(s)^{n_o \times n_i}$ has a normal rank r</u> s.t.

7 $$r = \mathrm{rk}\hat{H} = \min(n_o, n_i).$$

10 <u>Theorem</u> [Zeros, $n_i \leq n_o$]. Consider the transfer function matrix $\hat{H} \in \mathbb{R}(s)^{n_o \times n_i}$ where $n_i \leq n_o$ and (7) holds.

U.t.c.

If $z \in Z[\hat{H}]$, then $\exists$ a nonzero vector $\xi \in \mathbb{C}^{n_i}$ and $\hat{m}(s) \in \mathbb{C}[s]^{n_i}$ s.t. for the input given by

11 $$u(t) := \xi \exp(zt) + m(t) \quad \forall t \geq 0,$$

the output $\hat{y}(s) = \hat{H}(s)u(s)$ is given by

12 $$y(t) = n(t) \qquad \forall t \geq 0$$

with
$$\hat{n}(s) \in \mathbb{C}[s]^{n_o}.$$

13 <u>Comments</u>. (a) Note that both $m(\cdot)$ and $n(\cdot)$ are distributions with support at $t = 0$. Hence $\forall t > 0$, the input $u(t) = \xi \exp(zt)$, and it produces

an output $y(t) = \theta$, $\forall t > 0$. We observe the blocking property of the zero z. For this reason, z is often called a zero of transmission.

(b) The distribution $m(\cdot)$ "kicks" the system such that none of the modes associated with the possible poles of $\hat{y}(s)$ is excited.

14 Proof of Theorem 10. Let (D_ℓ, N_ℓ) be a ℓ.c.f. of $\hat{H}$, whence $\exists\, U_\ell, V_\ell \in E(\mathbb{R}[s])$ s.t.

15 $$N_\ell U_\ell + V_\ell D_\ell = I_{n_o}.$$

Now by assumption $z \in Z[\hat{H}]$, hence, by Theorem 2.4.4.6, $rk[N_\ell(z)] < n_i$, whence

16 $\exists$ a nonzero vector $\xi \in \mathbb{C}^{n_i}$ s.t. $N_\ell(z)\xi = \theta_{n_o}$.

Therefore, also by (16)

17 $$\hat{\zeta}(s) := N_\ell(s)\xi/(s - z) = (N_\ell(s) - N_\ell(z))\xi/(s - z) \in \mathbb{C}[s]^{n_o}.$$

Define now

18 $$\hat{m}(s) := -U_\ell(s)\,\hat{\zeta}(s) \in \mathbb{C}[s]^{n_i},$$

19 $$\hat{u}(s) := \xi/(s - z) + \hat{m}(s).$$

Then by (18)-(19), (17), and (15)

$$\begin{aligned}\hat{y}(s) = \hat{H}(s)\hat{u}(s) &= D_\ell(s)^{-1}N_\ell(s)\,\{\xi/(s - z) - U_\ell(s)\,\hat{\zeta}(s)\} \\ &= D_\ell(s)^{-1}\,\{I - N_\ell(s)\,U_\ell(s)\}\,\hat{\zeta}(s) \\ &= D_\ell(s)^{-1}\,D_\ell(s)\,V_\ell(s)\,\hat{\zeta}(s) \\ &= V_\ell(s)\,\hat{\zeta}(s) =: \hat{n}(s) \in \mathbb{C}[s]^{n_o}.\end{aligned}$$ ∎

22 Theorem [Zeros, $n_o \le n_i$]. Consider the transfer function matrix $\hat{H} \in \mathbb{R}(s)^{n_o \times n_i}$ where $n_o \le n_i$ and (7) holds.

U.t.c.

If $z \in Z[\hat{H}]$ and $z \notin P[\hat{H}]$, then $\exists$ a nonzero vector $\eta \in \mathbb{C}^{n_o}$ such that for every nonzero vector $\xi \in \mathbb{C}^{n_i}$ $\exists$ $\hat{m}_\xi(s) \in \mathbb{C}[s]^{n_i}$ s.t. the input

23 $$u(t) := \xi \exp(zt) + m_\xi(t) \qquad \forall t \geq 0$$

produces an output $\hat{y}(s) = \hat{H}(s)\hat{u}(s)$ satisfying

24 $$\eta^* y(t) = 0 \qquad \forall t \geq 0.$$

25 Comment. In contrast to Theorem 10, Theorem 22 allows any nonzero $\xi \in \mathbb{C}^{n_i}$ and asserts that the transmission of $\xi \exp(zt)$ has been blocked in the output direction defined by η: $\forall t > 0$ "$y(t)$ can only belong to the hyperplane orthogonal to η, viz. $< \eta, y(t) > = 0$."

26 Proof of Theorem 22. Let (D_ℓ, N_ℓ) be a ℓ.c.f. of H. Since $z \in Z[H]$, by Theorem 2.4.4.6, $rk[N_\ell(z)] < n_o$, and

27 $\exists$ a nonzero vector $\gamma \in \mathbb{C}^{n_o}$ s.t. $\gamma^* N_\ell(z) = \theta^*_{n_i}$. Since (D_ℓ, N_ℓ) is ℓ.c., $rk[D_\ell(z) \quad N_\ell(z)] = n_o$ and it follows from (27) that

28 $$\eta^* := \gamma^* D_\ell(z) \neq \theta^*_{n_o} .$$

For any nonzero $\xi \in \mathbb{C}^{n_i}$, let now

29 $$h_\xi(s) := \gamma^* D_\ell(z) D_\ell(s)^{-1} N_\ell(s)\xi \in \mathbb{C}(s).$$

Then by (27) and the fact that $z \notin P[H]$

30 $$h_\xi(z) = 0$$

s.t.

31 with (n,d) a coprime fraction of h_ξ, $n(z) = 0$, $d(z) \neq 0$.

Consequently, since n and d are polynomials,

32 $$\nu(s) := n(s)/(s - z) \in \mathbb{C}[s]$$

and

33 $$\exists\ \alpha \neq 0 \quad \text{s.t.} \quad -1 + \alpha d(z) = 0.$$

Hence by (33),

34 $$\pi(s) := (-1 + \alpha d(s))/(s - z) \in \mathbb{C}[s],$$

35 $$[1 + (s - z)\pi(s)]/d(s) \equiv \alpha \neq 0.$$

Using the ξ chosen above, define

36 $$\hat{m}_\xi(s) := \xi\pi(s) \in \mathbb{C}[s]^{n_i}$$

and

37 $$\hat{u}(s) := \xi/(s - z) + \hat{m}_\xi(s).$$

Then, using (28), (36)-(37), (29), (31), (32), and (35) it follows that

$$\begin{aligned}\eta^* \hat{y}(s) &= \gamma^* D_\ell(z)\, D_\ell(s)^{-1}\, N_\ell(s)\, \xi\,[(s - z)^{-1} + \pi(s)] \\ &= h_\xi(s)\,[(s - z)^{-1} + \pi(s)] \\ &= (n(s)/(s - z))\,\{[1 + (s - z)\pi(s)]/d(s)\} \\ &= \nu(s)\,\alpha \in \mathbb{C}[s].\end{aligned}$$ ∎

38 Exercise. Use the SMM-form equation $H = L\Psi_\ell^{-1}\mathcal{E}R$ to show that (30) follows from (29).

39 Comment. The dynamical interpretation of poles is fundamental to the understanding of stability questions (see Sec. 4.2). The dynamical interpretation of zeros is basic to the understanding of certain requirements imposed in the tracking problem (see Sec. 5.2). References for this section are [Des.1], [Cal.2].

2.5. Realization and Polynomial Matrix Fractions

This section is only a brief outline. For a complete treatment, see [Kai.1, Sec. 6.4, Sec. 6.5].

1.R Fact. Assume that we are given a PMD $[D_r, I, N_r, 0]$ corresponding to the equations

2 $$D_r(p)\xi(t) = u(t)$$

$$t \geq 0$$

3 $$y(t) = N_r(p)\xi(t)$$

and giving rise to a strictly proper transfer function satisfying

4 $$H = N_r D_r^{-1} \in \mathbb{R}_{p,o}(s)^{n_o \times n_i} \text{ s.t. } \underline{\text{rk } H = n_i}$$

U.t.c. there exists a controllable realization $[A, B, C, 0]$ of H having the SSD

5 $$px(t) = Ax(t) + Bu(t)$$

$$t \geq 0,$$

6 $$y(t) = Cx(t)$$

where $x(t) \in \mathbb{R}^n$, with

7 $$n = \partial[\det D].$$ ■

This is based on the following properties of (N_r, D_r):

(a) By Remark 2.4.3.14 and Theorem 2.3.3.18 we may assume without loss of generality that

8 $$D_r \in \mathbb{R}[p]^{n_i \times n_i} \text{ is column-reduced,}$$

s.t., if

9 $$k_j := \partial_{cj}[D_r] \qquad \forall j \in \underline{n}_i,$$

then

10 $$n := \partial[\det D] = \sum_{j=1}^{n_i} k_j.$$

(b) By (4) and Theorem 2.4.3.25.R,

11 $$0 \le \partial_{cj}[N_r] < \partial_{cj}[D_r] = k_j \qquad \forall j \in \underline{n}_i .$$

(c) Because of (8)-(11) and Fact 2.4.3.21.R, $(N_r, D_r) \in \mathbb{R}[p]^{n_o \times n_i} \times \mathbb{R}[p]^{n_i \times n_i}$ can be represented by

12 $$D_r(p) = D_h S(p) + D_\ell \Psi(p),$$

13 $$N_r(p) = N_\ell \Psi(p),$$

where

14 $$S(p) := \text{diag}[p^{k_j}]_{j=1}^{n_i} \in \mathbb{R}[p]^{n_i \times n_i},$$

15 $$\Psi(p) := \text{block diag}\left\{ \begin{bmatrix} p^{k_j-1} \\ \vdots \\ p \\ 1 \end{bmatrix}, k_j \times 1, j \in \underline{n}_i \right\} \in \mathbb{R}[p]^{n \times n_i},$$

16 $D_h \in \mathbb{R}^{n_i \times n_i}$ is the nonsingular highest column-degree coefficient matrix (2.4.3.15),

17 $D_\ell \in \mathbb{R}^{n_i \times n}$ is a coefficient matrix taking into account lower-degree terms of each entry of D_r,

18 $N_\ell \in \mathbb{R}^{n_o \times n}$ is a coefficient matrix taking into account the entries of N_r.

Procedure. Using (12)-(18), write equations (2)-(3) as

19 $$S(p)\xi(t) = -D_h^{-1} D_\ell \Psi(p) \xi(t) + D_h^{-1} u(t)$$

20 $$y(t) = N_\ell \Psi(p) \xi(t).$$

Hence a state $x(t) \in \mathbb{R}^n$ can be constructed as

21 $$x(t) := \Psi(p)\xi(t) = (p^{k_1-1}\xi_1(t), \cdots, p\xi_1(t), \xi_1(t); \cdots; p^{k_{n_i}-1}\xi_{n_i}(t), \cdots, p\xi_{n_i}(t), \xi_{n_i}(t))^T$$

using n_i integrator chains of length k_j driven by $p^{k_j}\xi_j(t)$, a first-order derivative of a component of the state $x(t)$ by (21). Note that by (19) and (21), $\forall j \in \underline{n}_i$, $p^{k_j}\xi_j(t)$ is an $\mathbb{R}$-linear combination of state and input components: this generates n_i equations of the state equation $px(t) = Ax(t) + Bu(t)$. This equation is then completed using $n - n_i$ integrator relations of the form $px_\ell(t) = x_{\ell-1}(t)$. The readout map $y(t) = Cx(t)$ follows from (20)-(21).

For example, let

$$D_r(p) = \begin{bmatrix} (p+1)^2 & p \\ 0 & p-1 \end{bmatrix}, \quad N_r(p) = \begin{bmatrix} p & 0 \\ 1 & 1 \end{bmatrix}.$$

Then $n_o = n_i = 2$, D_r is c.r., $k_1 = 2$, $k_2 = 1$, $n = k_1 + k_2 = 3$. (12)-(18) result in:

$$S(p) = \operatorname{diag}[p^2, p],$$

$$\Psi(p) = \left[\begin{array}{c|c} p & 0 \\ 1 & 0 \\ \hline 0 & 1 \end{array}\right],$$

$$D_h = \begin{bmatrix} 1 & 1 \\ 0 & 1 \end{bmatrix},$$

$$D_\ell = \left[\begin{array}{cc|c} 2 & 1 & 0 \\ \hline 0 & 0 & -1 \end{array}\right], \quad N_\ell = \left[\begin{array}{cc|c} 1 & 0 & 0 \\ \hline 0 & 1 & 1 \end{array}\right].$$

Hence (19) reads

$$p^2\xi_1(t) = -2p\xi_1(t) - \xi_1(t) - \xi_2(t) + u_1(t) - u_2(t),$$

$$p\,\xi_2(t) = \xi_2(t) + u_2(t).$$

The state is constructed by (21) as

$$x(t) = (x_1(t), x_2(t); x_3(t))^T := (p\xi_1(t), \xi_1(t); \xi_2(t))^T$$

using $n_i = 2$ integrator chains of length $k_1 = 2$, $k_2 = 1$ resp. and driven by $p^2\xi_1(t) = px_1(t)$, $p\xi_2(t) = px_3(t)$ resp. Using two equations generated by (19) and (21) and one integrator relation, the state equation $px(t) = Ax(t) + Bu(t)$ is s.t.

$$A = \left[\begin{array}{cc|c} -2 & -1 & -1 \\ 1 & 0 & 0 \\ \hline 0 & 0 & 1 \end{array}\right], \quad B = \left[\begin{array}{cc} 1 & -1 \\ 0 & 0 \\ \hline 0 & 1 \end{array}\right].$$

From (20)-(21) we have a readout map $y(t) = Cx(t)$ with

$$C = \left[\begin{array}{cc|c} 1 & 0 & 0 \\ \hline 0 & 1 & 1 \end{array}\right].$$

<u>Controllability</u>. Note that the state equation $px(t) = Ax(t) + Bu(t)$ is obtained from the state equation $px(t) = A^o x(t) + B^o v(t)$ of the $n_i = 2$ integrator chains described by

$$A^o = \left[\begin{array}{cc|c} 0 & 0 & 0 \\ 1 & 0 & 0 \\ \hline 0 & 0 & 0 \end{array}\right], \quad B^o = \left[\begin{array}{cc} 1 & 0 \\ 0 & 0 \\ \hline 0 & 1 \end{array}\right], \quad (A^o, B^o) \text{ controllable, through}$$

the application of a feedback law given by

$$v(t) = D_h^{-1}(u(t) - D_\ell x(t)).$$

The latter represents an input-coordinate transformation followed by state feedback. As a result

$$A = A^o - B^o D_h^{-1} D_\ell \qquad B = B^o D_h^{-1},$$

as is easily checked. Since (A^o, B^o) is controllable it follows that (A, B) is controllable because the transformations preserve controllability. Notice finally that the above realization is observable and therefore minimal iff (N_r, D_r) is r.c. [Kai.1, Th. 6.5 -1].

23 <u>Exercise</u>. Develop a dual technique to obtain an <u>observable</u> realization of dimension $n = \partial[\det D_\ell]$ for the PMD $[D_\ell, N_\ell, I, 0]$ giving rise to a transfer

function $H = D_\ell^{-1} N_\ell \in \mathbb{R}_{p,o}(s)^{n_o \times n_i}$ s.t. rk $H = n_o$. Call this Fact 1.L [Kai.1, pp. 413-417].

24 Exercise. Study the relation between coprime fractions and minimal realizations in [Kai.1, pp. 439-440].

Chapter 3. Polynomial Matrix System Descriptions and Related Transfer Functions

3.1. Introduction

This chapter studies the time-domain properties of a polynomial matrix system description (PMD) and related algebraic properties of transfer functions. Section 2 studies the dynamics of a PMD (pseudo-state trajectory, response, ···) and the possibility of dynamical redundancy (reachability, observability, hidden modes, ···). Section 3 investigates the quality of the behavior at $t = 0$ of a PMD (well-formed PMDs, ···), and the exponential stability of a PMD (exponentially decreasing pseudo-state trajectory, ···). Section 4 describes properties of transfer functions generated by a PMD: right-left fractions and internally proper fractions (corresponding to well-formed PMDs).

3.2. Dynamics of a PMD; Redundancy

3.2.1. Dynamics of a PMD

Consider the PMD $[D, N_\ell, N_r, K]$ described by the equations

1 $$D(p)\xi(t) = N_\ell(p)u(t)$$

$$t \geq 0$$

2 $$y(t) = N_r(p)\xi(t) + K(p)u(t),$$

where (a) $D(p) \in \mathbb{R}[p]^{\nu\times\nu}$, $N_\ell(p) \in \mathbb{R}[p]^{\nu\times n_i}$, $N_r(p) \in \mathbb{R}[p]^{n_o\times\nu}$, $K(p) \in \mathbb{R}[p]^{n_o\times n_i}$;

(b) $D(\cdot)$ is nonsingular;

(c) $u(\cdot)\colon \mathbb{R}_+ \to \mathbb{R}^{n_i}$, $\xi(\cdot)\colon \mathbb{R}_+ \to \mathbb{R}^{\nu}$, $y(\cdot)\colon \mathbb{R}_+ \to \mathbb{R}^{n_o}$ are called the <u>input</u>, <u>pseudo-state</u>, and <u>output</u> of the PMD;

(d) $n := \partial[\det D(\cdot)] > 0$ is called the <u>order</u> of the PMD;

(d) $u(\cdot)$ is piecewise sufficiently differentiable (see the definition below). Note that the presence of the differentiation operator $p = d/dt$ in $N_\ell(p)$ and $K(p)$ requires the knowledge of the value of $u(\cdot)$ and a sufficient number of its derivatives at $t = 0-$.

3 A function $f : \mathbb{R}_+ \to \mathbb{R}^n$ is said to be piecewise sufficiently differentiable (p. suff. diff.) iff (a), except for a set $D \subset \mathbb{R}_+$, $f \in C^r$ (equiv. f is r times continuously differentiable, where r is large enough so that all differentiations are well defined); (b) D contains at most a finite number of points per unit interval; (c) $\forall \tau \in D$, $\forall k = 0 \sim r$, $f^{(k)}(\tau-)$ and $f^{(k)}(\tau+)$ are well defined and finite.

4 A typical example of a p. suff. diff. function (with r = 0) is $t \mapsto k(1(t) - 1(t-1))$ with $k \in \mathbb{R}^n$ and $1(\cdot)$ the unit step function.

5 We call state at t = 0 of the PMD described by (1)-(2), any state $x(0) \in \mathbb{R}^n$ of the differential equation $D(p)\xi(t) = \theta_\nu$ $t \geq 0$ (see Definition 2.3.5.16).

6 We call zero-input pseudo-state trajectory (z-i p-s trajectory) of the PMD any C^∞-solution $\xi(\cdot) : (0-, \infty) \to \mathbb{R}^\nu$ of the differential equation $D(p)\xi(t) = \theta_\nu$ $t \geq 0$. Hence by definition $\xi^{(j)}(0-) = \xi^{(j)}(0+)$ $\forall j \in \mathbb{N}$.

7 Among the states at t = 0 of the PMD we note the (normalized) state $x_\xi(0) \in \mathbb{R}^n$ of the PMD, which is defined as the normalized state $x_\xi(0) \in \mathbb{R}^n$ of the differential equation $D(p)\xi(t) = \theta_\nu$ $t \geq 0$ (see Definition 2.3.5.43).

8 Note that by, Algorithm 2.3.5.36, $x_\xi(0) \in \mathbb{R}^n$ is given in terms of $\xi(0)$ and some of its derivatives. Note that by Theorem 2.3.5.38 and Definition 6 the z-i p-s trajectory of the PMD has the form

9 $$\xi(t) = \psi(t)\, P x_\xi(0) \qquad \forall t > 0-,$$

where $\Psi(\cdot)$ is any basis matrix (2.3.5.13) for the solution space X of the differential equation $D(p)\xi(t) = \theta_\nu$ $t \geq 0$, and $P \in \mathbb{R}^{n \times n}$ is a nonsingular matrix.

10 We call zero-input response (z-i response) of the PMD the response $y(\cdot) : \mathbb{R}_+ \to \mathbb{R}^{n_o}$ of the PMD when $u(t) \equiv \theta$ $\forall t > 0-$.

11 Note that by the substitution of (9) in (2) the z-i response of the PMD is given by

12 $$y(t) = N_r(p)\xi(t) = N_r(p)\psi(t)Px_\xi(0), \quad \forall t > 0-,$$

where differentiations are to be made in the usual sense because $\xi(\cdot)$ and all its derivatives are continuous on $t > 0-$.

15 We call <u>characteristic polynomial</u> of the PMD the polynomial det $D(\lambda)$, and <u>eigenvalue of the PMD</u> any $\lambda \in \mathbb{C}$ s.t. det $D(\lambda) = 0$.

16 By using Laplace Transform techniques it follows that any z-i p-s trajectory $\xi(\cdot)$ of the PMD is an sum of exponential polynomials in t of the form

17 $$\sum_{k=0}^{m} (a_k t^k)\exp(\lambda t) \quad \forall t > 0-,$$

where $a_k \in \mathbb{C}^\nu$ $\forall k = 0 \sim m$, $\lambda \in \mathbb{C}$ is an eigenvalue of the PMD, with the restriction that iff $\lambda \in \mathbb{C}\setminus\mathbb{R}$ then $\xi(\cdot)$ contains also the complex conjugate companion form

18 $$\sum_{k=0}^{m} (\bar{a}_k t^k)\exp(\bar{\lambda} t) \quad \forall t > 0-.$$

21 We call <u>zero-state pseudo-state trajectory</u> (z-s p-s trajectory) <u>of the PMD</u> the solution $\xi(\cdot)$ of the differential equation

1 $$D(p)\xi(t) = N_\ell(p)u(t) \quad t \geq 0$$

driven by a piecewise sufficiently differentiable input $u(\cdot) : \mathbb{R}_+ \to \mathbb{R}^{n_i}$ where both $\xi(\cdot)$ and $u(\cdot)$ and all their derivatives are zero at $t = 0-$ (more precisely $\xi^{(i)}(0-) = \theta_\nu$ and $u^{(j)}(0-) = \theta_{n_i}$ $\forall i = 0, 1, 2, \cdots$ and $\forall j = 0, 1, 2, \cdots$).

22 As a result the z-s p-s trajectory can be calculated as follows. Taking the Laplace transform on both sides, we obtain

23 $$\hat{\xi}(s) = D(s)^{-1}N_\ell(s)\hat{u}(s)$$

where by division Theorem 2.4.3.37.L